Governing Urban Development in China

The book investigates urban development and governance in China and introduces China perspectives to the understanding of governing urban development in the 21st century.

Building upon a rich and burgeoning literature on China, the book explains major changes in governance, offers a well-synthesized account of state-centered governance, and provides in-depth discussions on urban governance, city and regional planning, financing and financialization, urban redevelopment, local economic development and innovation, and environmental governance. The book bridges theoretical concepts in critical urban studies and empirical research on China and thus depicts a fuller picture of changing and variegated urban governance in the contemporary world. The book theorizes Chinese urban governance from the ground up and derives a concept of state entrepreneurialism as a framework for narrating urban governance in China. Following this framework, each chapter begins with a brief introduction to key concepts in urban geography and then depicts the urban development process on the ground in China. Then, the chapters discuss these concepts and explanations because many are derived from a different context, often in Western economies. At the end of each chapter, the phenomenal urban changes are evaluated with their theoretical implications.

This book offers contextualized insights into critical geographical studies of urban governance and is the first essential complementary reading for both urban scholars and those exploring the geography of China. It will be of interest to students and researchers in Urban Geography, Urban Studies, Urban Planning, Sociology, Political Science, and China Studies. The book can also be complementary reading in China Studies, especially in governance and politics.

Fulong Wu is Bartlett Professor of Planning at University College London. He received his BSc and MSc from Nanjing University, and his PhD from the University of Hong Kong. He has taught previously at Southampton and Cardiff universities. In 2016, he was conferred the award of Fellow of Academy of Social Sciences in the UK. His research interests include urban development in China and its social and sustainable challenges. He is the author of *Planning for Growth: Urban and Regional Planning in China* (Routledge, 2015), *Creating Chinese Urbanism: Urban Revolution and Governance Change* (UCL Press, 2022), and co-editor (with Roger Keil) of *After Suburbia: Urbanization in the 21st Century* (2022). He is the principal investigator of a European Research Council (ERC) Advanced Grant – Rethinking China's Urban Governance.

Fangzhu Zhang is Professor of China Planning and joint coordinator for the China Planning Research Group in the Bartlett School of Planning, University College London (UCL), UK. Her main research interests focus on innovation and governance, environmental planning and eco-city development, urban financialization, and urban village redevelopment in China. She has been involved in several research projects funded by the British Academy, ESRC (UK), and the EU. She is a co-editor of *Handbook on China's Urban Environmental Governance*. She is the founding Editor-in-Chief of *Transactions in Planning and Urban Research*. Currently, she is working on the ERC Advanced Grant research project—Rethinking China's Urban Governance.

Governing Urban Development in China

Critical Urban Studies

Fulong Wu and Fangzhu Zhang

Routledge
Taylor & Francis Group
LONDON AND NEW YORK

Designed cover image: Fulong Wu and Fangzhu Zhang

First published 2025
by Routledge
4 Park Square, Milton Park, Abingdon, Oxon OX14 4RN

and by Routledge
605 Third Avenue, New York, NY 10158

Routledge is an imprint of the Taylor & Francis Group, an informa business

British Library Cataloguing-in-Publication Data
A catalogue record for this book is available from the British Library

ISBN: 978-1-032-40893-4 (hbk)
ISBN: 978-1-032-40896-5 (pbk)
ISBN: 978-1-003-35517-5 (ebk)

DOI: 10.4324/9781003355175

Typeset in Times New Roman
by SPi Technologies India Pvt Ltd (Straive)

The publication of this book in Open Access is supported by the European Research Council (ERC) through the advanced grant *Rethinking China's Urban Governance* (No. 832845).

Funded by
the European Union

European Research Council
Established by the European Commission

Contents

Figures

Preface

The photo on this book's cover depicts the landscape of Tianfu New Area in Chengdu. We took this photo in November 2023. Five years earlier, in 2018, precisely at this position (see the figure on the next page), President Xi Jinping endorsed the concept of the 'park-city' as an exemplar of ecological civilization. The park-city slogan suggests that Chengdu would build the whole city into a park or 'building the city inside parks.' We choose this photo to illustrate a new era of governing urban development in China.

However, after we selected this photo, we discovered that the landscape in the cover image is surprisingly similar to the cover of an earlier book by Fulong Wu, *Planning for Growth* (2015). Indeed, both cover images are about carefully designed and landscaped new towns by the lakeside (see the figure on the next page), which suggests that there is a universal model of Chinese urban development. *Planning for Growth* serves a different purpose—illustrating entrepreneurial city planning practices under market-oriented reform. It seems to imply a neoliberal shift; on the ground, however, it suggests state entrepreneurialism—governance beyond deregulation. It shows what planners have done instead of a streamlined planning system. However, the governance of urban development is not fully examined there. This book picks up the under-studied issue of urban governance and rethinks the role of the state in urban development.

What has changed in China's urban governance? We interrogate this question through a grounded observation, taking the same perspective as the Chinese top leader who observed the site before these buildings were completed. Here, we intend to use China as a laboratory to observe contemporary changes—a wider trend in present-day urban governance. These observations are contextually particular. However, we strive to avoid China's 'exceptionalism,' often implied in party-state authoritarianism.

For that purpose, we will discuss the literature on critical urban studies. From there, we show a logic consistent with what David Harvey has observed at the waterfront of Baltimore's Inner Harbour. Interestingly, researchers pay more attention to his finding—'from managerialism to entrepreneurialism' and forget that his seminal paper has a subtitle—'the transformation in urban governance in late capitalism,' which opens up a field of research.

Our book is situated in this field of research but from a new observation site, as shown in the cover image. Here, we characterize the transformation in urban governance as a shift to state entrepreneurialism, with explicit attention to capital accumulation and territorial politics. The book provides historical and geographical accounts of China's changing development and governance.

This book originated from making videos for teaching during the COVID-19 pandemic. No scripts were written at that time. In retrospect, it would have been easier if we had written the script before making amateur videos. Nevertheless, the exercise forced us to take a visual and grounded approach to the topics that were quite abstract to students.

(a)

Figure 0.1a The observation site at Tianfu New Area, Chengdu, where the 'park-city' concept was first mentioned during President Xi Jinping's visit in 2018.

(b)

Figure 0.1b The precise standing point is marked on the ground to commemorate President Xi Jinping's visit in 2018.

Figure 0.2 The Taihu New Town of Wuxi is located by man-made wetlands and lakes. The new municipal government is housed in a complex of two tower buildings.

Acknowledgements

The book grew out of our research under the European Research Council (ERC) advanced project Rethinking China's Urban Governance and our teaching on the Urban Planning in China course at University College London. However, our understanding of China's urban governance has been accumulated over a long time, and we are grateful for all the help and suggestions we have received from many colleagues over the years.

We are indebted to those who offered generous advice in research and teaching: Gonghao Cui, Anthony Gar-On Yeh, Chris Webster, John Logan, Laurence Ma, Yixin Zhou, Cecilia Wong, Phil Cooke, Yaping Wang, Jenny Robinson, Alan Penn, Nick Gallent, Mike Raco, Yvonne Rydin, Nick Phelps, and Le-Yin Zhang.

Thanks also go to those who have been companions in urban China research: Chaolin Gu, Yuemin Ning, Cindy Fan, You-tien Hsing, Weiping Wu, George Lin, Mark Frazier, Youqing Huang, Hyun Bang Shin, Xuefei Ren, Helen Bao, Canfei He, Zhi Liu, Julie Ren, Jiang Xu, and Yawei Chen.

We want to thank those who inspired us in critical urban studies. We are enlightened by your insights: Jenny Robinson, Nick Phelps, Roger Keil, Neil Brenner, Jamie Peck, Ananya Roy, Patrick Le Galès, Manuel Aalbers, Phil Harrison, Christian Schmid, Loraine Kennedy, Allan Cochrane, Andrew Jonas, Chris Hamnett, Henry Yeung, Richard LeGates, and Andy Pike.

We specifically thank those who work with us over the years; thank you for your support during various stages of fieldwork and research: Zhigang Li, Shenjing He, Yuting Liu, Jian Feng, Jie Shen, Yi Li, Tingting Lu, Zhen Wang, Jingxiang Zhang, Fenghua Pan, and Jie Chen, Calvin Chung, Siyao Liu, Manqi Wang, and Kan Zhu.

In addition, we particularly wish to thank those for their hospitality and help during our fieldwork: Yuan Yuan, Yang Xiao, Xiang Luo, Wen Chen, Lan Zhou, Desheng Xue, Pengjun Zhao, Sainan Lin, Xuemei Zhu, Juan Sun, Xiaodong Shi, Song Shi, Xiaohui Chen, Guofang Zhai, Mengzhu Zhang, Qiang Wang, Yuqi Liu and Ye Liu, Wenjing Mo and Fei Duan, Yanjing Zhao, Shuping Cui, Jinlong Liu, Wei Duan, Xingyun Mu, Yi Chen, Yong Wan and Linhui Wang, Xiangming Ma, Hongyang Wang, Xiaoming Kuang, Sougui Luo, Yun Qian, Yonghua Zou, Yumin Ye, Hai Yu, Hongou Zhang, Zeng Gang, Zhi Qian, Jiawen Yang, and Le Yu.

At UCL, we would like to thank our postdoctoral ERC team, especially Ying Wang, Weikai Wang, Yi Feng, and Zhenfa Li. In particular, Yi Feng, Zhenfa Li, Handuo Deng, Ying Wang, Manqi Wang and Jingyi Qin have read some of the initial chapters.

In addition, Chris Hamnet offered valuable revision suggestions. We thank three anonymous reviewers for their helpful and detailed suggestions and the commissioning editor Faye Leerink for efficient publication arrangement.

Chapter 1 was presented in a public lecture at the University of Hong Kong, Urban China Research Network Annual Conference, RSA China Conference, and its Chinese version in a public lecture at Wuhan University in 2024.

Finally, we acknowledge the support from the European Research Council (ERC) for funding open access under the advanced grant *Rethinking China's Urban Governance* (No. 832845).

Introduction

Is there a distinctive model of urban development in China? How are Chinese cities governed? What are the features of China's urban governance? These are empirical questions, but they also have significant theoretical implications. Beyond these empirical questions, this book offers a Chinese perspective on the understanding of governing urban development in the 21st century. It investigates new practices of governing Chinese cities. The book is not a reference for specialists in China. Rather, it is an intersection between urban studies and China studies. Situated in the conversations of critical urban studies, the book does not apply existing concepts to China. It uniquely combines interrogations of Chinese practices and reflections on the theories and concepts of critical urban studies. It reveals variegated approaches to urban development and governance. The book rethinks the exact meaning of these key concepts developed in Western economies and questions how they might be deployed to explain the phenomena in wider global urban studies (Robinson 2022).

At the crossroads of Chinese development and critical urban studies

The book does not provide a comprehensive view of Chinese cities. There is now a rich collection of urban China studies.[1] While paying attention to contextual specificities, the book provides a holistic historical and geographical account and develops an accessible narrative. It synthesizes burgeoning empirical studies on Chinese urban development and governance practices. Engaging with some key concepts of critical urban studies such as urban entrepreneurialism, neoliberal planning, financialization, property-led redevelopment, regional innovation system, and the socio-ecological fix, the book enriches them in a new context of Chinese urbanization.

Thus, the book illustrates the conceptual potential and limitations for geography and urban studies students who might be familiar with other contexts. It is not intended to be primarily theoretical, and the treatment of core concepts is basic and simplified.[2] The selection of concepts is partial and convenient, depending on our thematic purposes. Hence, theoretical engagement is not a systematic mapping exercise demonstrating a broad spectrum of theories.

Our purpose here is to add a new narrative to the urban imagination embedded in the West (Wu 2020a). Although this book does not directly compare East and West or North and South, it has a comparative background (Robinson 2022). Urban studies in China have emphasized China's uniqueness (Ren 2023). The post-reform Chinese urbanization experience is 'certainly not easily subsumed into standard discussions about urban development and urban change' (Hamnett 2020, 690). For example, the political dimension of urbanization is

DOI: 10.4324/9781003355175-1

salient—government policies heavily influence the urbanization process. However, emphasizing the state or 'embedded statism' may produce exceptional thinking (Teo et al. 2023).

How do we avoid Chinese exceptionalism (Ren 2023)? Stressing the role of the state in urban governance does not necessarily lead to exceptionalism (Wu 2023c). In contrast, governance in China demonstrates similarities to that of the Western cities and city-regions (Jonas 2020). The role of the state is formed along with China's conjunctural interaction with global geopolitics. Hence, we should theorize from the ground up (Robinson 2022). The approach taken in this book is a grounded understanding of China's practices of governing urban development—through a historical and geographical account. We review China's post-reform governance similarly to how neoliberalism is studied as a state project of 'class restoration' by David Harvey (2007) and 'state capitalism' is understood more recently (Alami and Dixon 2020).[3] These existing concepts offer a helpful perspective, an initial entry point of inquiry, and a narrative structure. Beyond exceptionalism thinking, the book situates empirical inquiries within a broader theoretical conversation.

We avoid distractive details and dialogue with theoretical concepts when confronted with rich empirical materials.

The book theorizes Chinese urban governance from the ground up and derives a concept of state entrepreneurialism as a framework for narrating urban governance in China. Following this framework, each chapter begins with a brief introduction to key concepts in urban geography and then depicts the urban development process on the ground in China. Then, the chapters discuss these concepts and explanations because many are derived from a different context, often in Western economies. They must be critically scrutinized with a more precise, subtle, and grounded understanding. At the end of each chapter, the phenomenal urban changes are evaluated with their theoretical implications.

The themes in this book are not selected because of the available theoretical literature. Instead, they are currently topical issues of urban development and governance in China, and these topics are chosen because of their practical relevance. Therefore, the research behind this book is purposeful rather than theoretically driven. Because of the balance between theoretical engagement and empirical coverage, the existing literature on critical urban studies is used only as an organizational device to narrate complex empirical findings from China.

The book bridges theories derived from different contexts by finding theoretical explanations of governance practices in China and their implications for theories. Thus, it makes perplexing Chinese practices more comprehensible to those who do not specialize in China. For urban scholars, it offers a different example and contextual challenge; for Chinese scholars, it provides a possible explanation of the logic of governance. The book is a complementary reading for urban studies, urban planning, and urban geography students.

The logic of governance

The core concern of this book about China's urban governance is the logic of governance. To understand how China is governed, attention has been focused on the government, especially various forms of authoritarianism. Beyond the neo-institutionalist concepts, such as adaptive and resilient authoritarianism, the evolving practices of power are examined (Shue and Thornton 2017). This perspective of governmentality investigates how power is executed. It examines government leadership, local states, and governance techniques applied to individuals. Through power practices, it hopes to uncover the logic of governance. Particularly relevant to urban governance is how the state acts at the grassroots level, involving bargaining and negotiation and applying the 'zoning' technique to govern state spaces.

Further, the logic of governance is studied by focusing more exclusively on the state bureaucracy system (Zhou 2022). It aims to identify the organizational logic of governance by investigating state practices such as bureaucratic muddling through collusive behaviors of the local states and collective action based on unorganized interests. The state apparatus presents a more complex picture than the 'entrepreneurial state' or the growth coalition for revenue maximization. According to this organizational logic, the central–local interaction produces high elite discretionary. The explanation seems closer to a regime analysis that focuses on political actors' concrete motivations derived from the structure of organizations. In the case of China, such an organization is centered upon the party-state. According to this Weberian perspective, the motivation is state self-interests for the benefit of politicians.

Similar to governmentality and the institutional analysis of state bureaucracy, this book pays attention to the role of the state (Wu and Zhang 2022). This role is understood from a political economy point of view. The state aims to construct and maintain the 'structural coherence' of the accumulation regime, and, hence, the particular form of governance is out of 'state strategic selectivity.' Structural coherence means a relatively stable and coherent structure within which capital accumulation can proceed, and the state's strategic selectivity indicates that to achieve such coherence, particular aspects of state capacities are selected as a 'strategy,' which includes 'spatial rescaling'—selecting a specific scale of state (for these concepts, see Chapter 1 for further explanation).

Thus, the logic of governance requires understanding the relationship between the state and entrepreneurialism—outside the government. China presents a contradictory picture of state authoritarianism and market prevalence. All theorists find that this combinational feature defies being categorized. They have to apply a denomination though 'with Chinese characteristics,' for example, 'neoliberalism with Chinese characteristics' (Harvey 2005), 'capitalism with Chinese characteristics' (Huang 2008), or Chinese-style state capitalism (Pearson et al. 2023).[4] From observing rich practices, the explanation is often selective. Scholars in the West from different theoretical stances can use China as a good example to support their theoretical intention.

Critical urban studies in the West, which emphasizes neoliberalism globally, tend to see China as an example of neoliberalism. In urban development, a concrete form is 'urban entrepreneurialism' (Harvey 1989). China specialists, on the other hand, tend to focus on the party-state (Pearson et al. 2023). They focus on key politicians, state politics, and policies. China's urban governance indeed demonstrates perplexing features. It presents a picture of intertwined state policies, market exchanges, and social agencies in everyday urban life. What is the logic of governing urban development in China?

State entrepreneurialism

The overarching framework for this book is state entrepreneurialism, which will be elaborated on in Chapter 2. State entrepreneurialism is a particular form of governance that deploys market mechanisms to achieve a political strategy that benefits class interests. It depicts a more universal rising market in advanced capitalism, as captured by neoliberalism. However, it emphasizes the role of the state in the tendency of market dominance and the impossibility of a self-regulating market.

State entrepreneurialism captures such a logic of governance. In China's urban studies, the logic is presented as 'planning centrality, market instruments' (Wu 2018b). The notion hopes to reconcile the conceptual tension between strong states and prevalent markets. More specifically, the narrative tries to theorize urban development in China from the ground up

(Wu 2023c). It understands the logic from the evolving historical process and conjunctural moments in which China is related to global geopolitics in the late capitalist world. It situates the logic in the process of capital accumulation and the political strategies derived from this process.

State entrepreneurialism responds to capitalism's challenge and internal political economy contradictions. Maintaining capital accumulation strengthens state legitimacy—the justification for staying in power. The pervasive use of market development has been utilitarianism, as shown in the late Qing dynasty's learning from the West. Capital accumulation was under severe constraints after 1949 and only revived half-heartedly in post-reform China. This partiality reflects the consistent emphasis on ideological and institutional centrality represented by the state apparatus.

However, marketization in China constantly challenges the traditional governance mode —state embedded in society (Wu 2022a) and authoritarianism organically established in a Confucian society. A modernization process has responded to the profound challenge through which the state is professionalized. It hopes to enhance state legitimacy eventually, and, as such, the strong and visible state is an inevitable outcome of maintaining 'structural coherence'—a coherent structure for capital accumulation.[5]

As such, marketization does not dislocate state centrality. Since Deng Xiaoping's 'Southern Tour' in 1992, a full-fledged market mechanism has been introduced. Paradoxically, this was initiated as a necessary countermeasure to solve the problems created by earlier market reform because further reform helps maintain the regime's stability. Marketization diverted some pressure on the state, as illustrated by the governmentality perspective on neoliberalism as a mobile technology (Ong 2007, 2011) and a tactic of 'class restoration' (Harvey 2007). In other words, neoliberalism is only a governance approach to achieve class interests, and, in capitalism, the class interests closely match capital accumulation. However, the concrete form of this governance can vary according to different geographical contexts and historical moments.

As governance at a distance often fails in the context of the Chinese modernizing society, the governance approach—marketization—calls for more direct state intervention. Recently, China has promoted neighborhood participation. However, participation is encouraged and organized by the state. It demonstrates an attempt to extend the party-state to the grassroots. This movement is predictable from a historical view of China's urban governance. The rising state is rooted in urbanization because the foundation of an 'earth-bound' society featuring rural China has been dismantled. Urbanization does not lead to urbanism outside the state's control. The state-centered governance is born out of state-led rebuilding of residential communities (Wu 2022a).

This logic to govern is described as 'state entrepreneurialism.' Seemingly, it contrasts with 'urban entrepreneurialism' coined initially as reactionary policies—reaction toward capital mobility—toward flexible accumulation. It is a similar tactical response to globalization, geopolitics, and the contradiction inherent in capital accumulation. A closer reading of Harvey's usage of urban entrepreneurialism and flexible accumulation reveals that they do not form a causal relationship for him. Urban entrepreneurialism is just equivalent to the tactics of flexible accumulation for class restoration.[6]

State entrepreneurialism follows a similar tactic, as both forms are not intended solely for capital accumulation. They can be seen in a more explicit treatment of the territorial interests through 'accumulation by dispossession' and destructive geopolitics in new imperialism (Harvey 2003). Similarly, state entrepreneurialism presents a similar underlying intentionality beyond capital accumulation. The differences between various forms of entrepreneurialism—more neoliberal or state-centered—are mainly due to their territorial and class structures. In China, a capitalist class has not fully formed. China is still at the frontier of capitalism—meaning it has not achieved such a status. Here, the interpretation might differ slightly from state capitalism as the latter believes state ownership represents a new de facto capitalist class. This difference has been identified in the literature on state capitalism (Alami et al. 2022; Alami 2023; Peck 2023), which tries to depict state capitalism through state ownership, in contrast to the 'free market economy' constituted by individual capitalists, as a source of explanation for a new form of more authoritarian governance.

Here, we need to understand China's specificity as it is peripheral to developed capitalist economies. Capital accumulation is often considered universal for capitalism, while territorial politics is believed to be specific to a national or local territory. The latter is geographically particular. Hence, explanations relying on territorial politics are often associated with exceptionalism—unique local actors in a local process. In other words, capital accumulation defines the structure, while territorial politics represents agencies and state actions. A broad spectrum of theoretical positions is produced depending on how these two aspects are treated (Deng 2023).

State entrepreneurialism represents a more structured explanation of the political economy within this spectrum. It does not regard China's governance as exceptional despite being an outlier (Hamnett 2020). The perspective does not rely on the agencies to explain China's governance changes. This does not deny that an actor's key decision at a particular historical moment—Deng Xiaoping's Southern Tour—can be considered a causal pathway. Neither is the everyday life of ordinary people irrelevant. However, these actors are situated in critical historical conjuncture and moments and the structures that include class relations in production and social reproduction. Hence, state entrepreneurialism seeks an explanation in totality beyond exceptionalism.

The structure of the book

This introductory chapter explains our intention to use China as a laboratory to observe contemporary changes in urban governance. It also introduces methodological considerations from comparative urban studies. The book highlights the importance of contextual particularities. At the same time, through a dialogue between critical urban studies and China research, it avoids the trap of 'exceptionalism' (Pow 2012; Ren 2023; Teo et al. 2023). The chapter also discusses the recent debate over urban governance in critical geographical studies.

Following this brief introduction, Chapter 1 introduces the concept of governance and various theoretical perspectives. It discusses the debate over neoliberalism and recent research on the shifts after austerity. The chapter highlights the insights from the Global South and reviews the extensive literature on China's governance. It then briefly introduces the concept of state entrepreneurialism through understanding governance as a mode of regulation over the contradiction and challenges in the political economy.

Chapter 2 further introduces the historical formation of state entrepreneurialism in China. It discusses some defining features and provides an overall framework for this book. As a framework, it treats the state and entrepreneurialism simultaneously and in a dialectic and combinational way. The chapter then illustrates it with examples of large-scale urban development projects. It critically reflects on the role of the state in governing urban regeneration, suburban development, and rural revitalization. These examples reveal some characteristics of

Chinese urban governance. Regarding recent attention to statecraft (Lauermann 2018; Pike et al. 2019; Thompson 2020, 2021; Roth et al. 2023), the chapter illustrates how mega-urban projects like new towns are conducted under the Chinese statecraft of state entrepreneurialism. The level of state intervention is arguably exceptional but indicates the state's role to varying degrees in different countries.

Following this framework, the rest of this book, beginning with Chapter 3, examines various governance aspects, including planning, financing and financialization, redevelopment, innovation governance, and environmental governance. Chapter 3 focuses on planning as a regulatory exercise and as a form of development politics. The chapter critically reflects on 'planning for growth' (Wu 2015b), which departs from reactionary governance to entrepreneurial statecraft. In this context, the exact meaning of post-political planning is reflected upon.

Moving on to financing urban development, Chapter 4 points out a widely known investment-driven approach in Chinese economic governance. It asks to what extent land-based finance resembles financialization. The chapter examines the historical change in housing commodification, urban development corporations, and local government debts. It reveals that rather than the evolving process of capital accumulation, financialization is triggered by the crisis of capital accumulation, specifically the financial crisis. China's financialization started from the state's effort to cope with the global financial crisis and deploy local government financial vehicles to finance urban development. In turn, the imperative came from the need to refinance local debts—as a statecraft managing financial crises. It is an outcome of state action rather than an initiative of the finance sector. The shift from financing development to financializing the debt reveals a long shadow of the state (Wu 2023a).

Chapter 5 then looks into concrete practices of urban redevelopment. A set of concepts derived from the West already depicts a picture of property-led redevelopment, culture-led regeneration, financialization, displacement, and gentrification. The chapter investigates the evolution of redevelopment practices in China and reveals how redevelopment has been prioritized, historically and now. It examines two types of redevelopment practices, i.e., informal settlement (urban villages) redevelopment and recent 'micro-regeneration.' They both illustrate state entrepreneurialism, which is beyond a local growth machine dynamic and reflects territorial logic. The urban redevelopment practice in China reveals the national political strategies (Wu et al. 2022b). From this, the chapter reflects on the earlier conceptualization of 'redevelopment as the frontier of neo-liberalization' (He and Wu 2009) and finds that so-called property-led regeneration and micro-regeneration present startlingly different forms but the same logic of governance.

State entrepreneurialism is a governance approach that fulfills state development strategies. This is illustrated well in the governance of innovation in Chapter 6. Focusing on innovation, economic geographers adopted a more relational approach beyond a locality view, with appropriate attention to global production networks (GPNs) (Coe and Yeung 2015). A rich literature, such as the coupling and decoupling of GPNs and regional economies, reveals both the role of networks and territorial assets (MacKinnon 2012). Although the theory insists it has comprehensively covered both firms and the state, the current literature pays more attention to the firms and their interconnections than the state and political economy specificities (Zhang and Wu 2019). However, understanding the innovation landscape requires a more explicit treatment of the state, its strategies, and its operations governing innovation. Neoliberalism is questioned for its explanatory power (Yeung 2019b), as a landscape of innovation can be attributed to neither the developmental state nor its absence. Examining

China's innovation spaces, the chapter stresses in agreement that China does not present a distinctive model of national innovation or its regional varieties. It is a hybrid model combining the state and market. Through the political economy and conjunctural analysis, China's 'national indigenous innovation' model is understood as using market means to achieve state strategic innovation objectives. This helps explain the presence and absence of innovation and other economic and geographical factors. State entrepreneurialism—the statecraft of economic innovation in this context—is not virtual, whereas the production network is considered a material causal condition.

Going beyond economic rationality, China's environmental governance illustrates the state's extra-economic functionality. Its environmental management lacks effective implementation. Greenwashing has been long noted in China. However, environmental governance has recently seen a green turn with the discourse of 'ecological civilization.' The literature on the urban sustainability transition, perceived from a multi-level perspective, aims to understand how environmental innovation has become a mainstream practice (Bulkeley et al., 2010). The perspective is insightful but generic — seemingly widely applicable to diverse contexts. However, attention to the Chinese context is much needed, and a cultural critique suggests the sustainability transition should consider local institutions and cultural traditions (Huang et al. 2021). The contextual understanding should include the political economy, including the role of the state. Chapter 7 examines the development of eco-cities and greenways in China. The chapter shows that the concept of the socio-ecological fix demonstrates its value because it explains the logic of the multi-scalar state in tackling environmental crises. It also helps reveal the state's intentionality—'ecological civilization'—in its environmental governance.

This book employs state entrepreneurialism to elucidate the logic behind governing urban development in contemporary China. It pays attention to the context of urban development. The key feature of China's urban governance is the role of the state in governing urban development, as shown in this book across city planning, development finance, urban regeneration, economic innovation, and environmental transition. In many ways, the China perspective contrasts with a more neoliberal approach.

Nevertheless, the context means more than a geographical difference. Hopefully, it will be a 'reflexive theorizing with socially engaged inquiry and context-rich, historicised modes of analysis' (Peck 2023, p. 1). Through the dialogue with critical urban studies, this book insists on not treating China as an exception, as state entrepreneurialism offers a perspective to stress that capital accumulation and territorial politics co-exist in urban governance. In addition, as illustrated in governing urban development, the perspective shows the specific governance form in which these two aspects are combined, in particular, the actual existence of statecraft—for the government to stay in office and for the political party to maintain its reign.[7] Because managing the city, unlike a firm, cannot be entirely executed based on property ownership, governing by market institutions is ultimately an explicitly resisted ideology, impossible to adopt, and practically not attempted in China. China is mainly pre-modern and on the trajectory toward modernization, producing a more visible state (Wu 2022a).

Like recently acknowledged 'municipal entrepreneurialism' in those market economies (Lauermann 2018; Robinson et al. 2022), the view of state entrepreneurialism explains how the state maintains its strategic and extra-economic intention through deploying and mobilizing market and society — to create its agents and to co-opt those that are already existent or emerging (Wu et al. 2024). Chinese examples expand the scope of inquiry and enrich critical urban studies concepts. They helps to present a fuller picture of governing urban development in the 21st century.

Further readings

Introductory texts include *China's Urban Transition* (Friedmann 2005), *The Chinese City* (Wu and Gaubatz 2012, 2nd edition 2021), *Urban China* (Ren 2013), and *Creating Chinese Urbanism* (Wu 2022a). The last book provides mainly an urban social geography perspective on Chinese neighborhoods, urban transformation, and grassroots governance.

In addition, there are edited volumes such as *Restructuring the Chinese City* (Ma and Wu 2005), *China's Emerging Cities* (Wu 2007), *Urban China in Transition* (Logan 2008), *The City in China: New Perspectives on Contemporary Urbanism* (Forrest et al., 2019), *Chinese Cities in the 21st Century* (Huang 2020), and, more recently, *China Urbanizing* (Wu and Gao 2022).

There are also more specialized handbooks, including *Handbook on Urban Development in China* (Yep et al. 2019), *Handbook on China's Urban Environmental Governance* (Zhang and Wu 2023), and *Handbook on Local Governance in China: Structures, Variations, and Innovations* (Ergenc and Goodman 2023).

Specifically, on China's governance, *Governance and Politics of China* is now a classic (4th edition, Saich 2015). *The State and Capitalism in China* (Pearson et al. 2023) is a short book on the political economy of China, focusing on the party and state capitalism.

An Urban History of China (Lincoln 2021) provides an accessible account of the historical development of Chinese cities. *Chinese Urbanism: Critical Perspectives* (Jayne 2018) is a cultural and post-structural overview of urban life in China. *Property Rights and Urban Transformation in China* (Qian 2022) examines urban development in China from the perspective of land property rights.

For the theories and concepts of governance and urban changes, there are plenty of texts and in-depth studies, for example, *Understanding Urban Policy* (Cochrane 2007), *The Politics of Urban Governance* (Pierre 2011), *The Urbanism of Exception* (Murray 2017), and *New Urban Spaces* (Brenner 2019), and more specialized handbooks on governance and state politics, such as *The Routledge Handbook on Spaces of Urban Politics* (Ward et al. 2018), and *Handbook on the Changing Geographies of the State* (Moisio et al. 2020). In addition, *Urban Theory: New Critical Perspectives* (Jayne and Ward 2016) covers key concepts in critical urban studies, including those directly relevant to this book, such as 'governance,' 'neoliberalism,' and 'politics.'

Notes

1 See further readings at the end of this chapter.
2 See further readings at the end of this chapter for some introduction to urban governance and urban transformation.
3 Some studies also view China as state capitalism, see Naughton and Tsai (2015).
4 The combination could be in a different contradictory way, 'socialism with Chinese characteristics' (Lim 2014).
5 See Jessop (2006, 2016) for the term of structured coherence. It is derived from an understanding of the economy's operation that is fundamentally incomplete and contradictory, leading to the structural crisis. Therefore, new institutional changes are needed to establish temporal coherence. This understanding comes from the regulation school.
6 This becomes clearer when he described the territorial interest in the new imperialism.
7 Noted examples are Peck's (2023) greater attention to the party-state and David Harvey's shift from neoliberal capitalism to a variant of socialism in China (see Peck 2023, p. 10).

1 Governance

Theories and perspectives

Introduction

Since this book examines governing urban development in China, we will first define the key concept of 'governance' and its variegated forms. The interest in 'governance' emerged in the 1990s and originated from the notion of 'governing without government' (Rhodes 1996). Unlike government, governance 'looks at the interplay between state and society and the extent to which collective projects can be achieved through a joint public and private mobilization of resources' (Pierre 2011, p. 5). This means that the public sector, namely the government, does not have to deliver all services. The direction of change is described as a shift from government to governance. Governing is achieved through public entrepreneurship, public–private partnerships, and self-governing networks (Rhodes 1996).

However, this alleged shift is conceptualized in association with an ideology of neoliberalism rather than an 'actually existing' one. Outside the UK, where the notion of governance originated, there has been a long tradition of interactions between the state and societal actors outside the government. It is argued that what has changed is the role of government in the governance process (Pierre 2011; Pierre and Peters 2020). This raises attention to how the state governs, for example, urban development through policies centered upon urban development corporations (Imrie and Thomas 1999) or managing mega-urban projects through market contracts (Raco 2014) and, more broadly, the governance techniques of 'post-politics' (Swyngedouw 2009). The government continues to play a key role in governance. So, the meaning of 'governance' may not be limited to its implied minimalist or regulatory state, but could include a wide spectrum from state-centric governance to 'governing without government' (Pierre 2011, p. 9). With different types of societal involvement, there are models of managerial governance associated with a 'new public management' (Phelps and Miao 2020),[1] corporatist governance, pro-growth governance, and welfare governance (Pierre 2011).[2]

In contrast to the notion of governing without government, this book investigates the role of the state in governance. Here, the concept of governance does not take the shift for granted. We situate the practices of governing Chinese cities in the political economy. This chapter still uses the concept of governance to examine the complex relationship between state, market, and society in urban development.

Here, we characterize Chinese urban governance as 'state entrepreneurialism.' This is defined for our purposes as a 'series of state entrepreneurial actions to fulfill its strategic intention to maintain economic growth, stability, and capital accumulation and in turn its governance capacity' (Wu 2023b, p. 365). In short, the concept describes a dialectic relationship between the state and market. In China, the state is above market principles—as a long history of non-capitalist economy. At the same time, however, market exchange is becoming

DOI: 10.4324/9781003355175-2

increasingly important, replacing reciprocal and redistributive principles for economic coordination and social integration (Wu 2022a).

The elevation of market rationality paradoxically reflects that the state is above market principles rather than vice versa, as the state sets the boundary within which market exchange can work at specific historical moments. The expansion of market rationality often occurs when China, as a non-capitalist economy, encounters external threats and internal crises. Like neoliberalism, the market is perceived as a solution, as 'there is no alternative.' Here, the perspective of state entrepreneurialism, intuitively regarded as a strong state, exposes the constraint to the state.

In short, this book describes how the state governs urban development through 'entrepreneurial statecraft'—the art of maintaining state legitimacy through innovations. Here, it is important to distinguish between intention and action. Also, it is important to distinguish intention and consequences, as they might not be expected. For example, the intention to promote the objective of 'ecological civilization' may not eventually lead to more ecological development, especially if the local governments reinterpret the rationality to suit their interests.

The purpose of governing is to maintain a 'structural coherence' because the political economy, driven by its underlying dynamic of capital accumulation, tends to run into periodic crises. The term 'structural coherence' refers to the coherence of the political economy structure (Jessop 2002a). Because of the contradiction of capital accumulation, its periodical crises indicate the lack of 'structural coherence' at a particular moment. A stable regime is then temporarily set up based on coherence. The term is used by Jessop (2002a) to suggest the role of the state in fixing and providing structural coherence: 'It claims that what is tendentially replacing the Keynesian welfare national state is the Schumpeterian workfare postnational regime. I suggest how this new form of state could contribute to the structural coherence of a new spatio-temporal fix for capital accumulation' (p. 10), and 'this social fix helps secure a relatively durable structural coherence in managing the contradictions and dilemmas inherent in the capital relation, so that different forms, institutions and practices tend to be mutually reinforcing' (p. 48).

The state's actions are versatile, aiming to achieve its intention. At historical moments of globalization and also at the end of the Cold War, actions through and in the market became possible policy choices. However, structural coherence is an objective desired by the state. Here, maintaining structural coherence is an integral objective of the territorial logic—such coherence is achieved in a territorial form or within a territory or state space. Internal contradictions and external geopolitical changes constantly challenge it, and it is not entirely achievable. Therefore, the actual governance approaches change along with the changing political economy.

This variegated form of governance—state entrepreneurialism—reflects a combination of 'planning centrality and market instruments' (Wu 2018b). This book will explain what has precisely changed in the statecraft, parallel to the 'scalar' change beyond a single nation-state taking full charge (Brenner 2004, 2019). This naturally invokes multi-level and multi-scalar thinking, and even a network or assemblage perspective of the state (Allen and Cochrane 2010). An understanding of central–local relations is essential in studying China's governance. While state entrepreneurialism does not exclude such a way of thinking, the essence of state entrepreneurialism emphasizes state rationality as a whole.

It seems that emphasizing the role of the state might lead to exceptionalism because China's visible state's role in governing urban development is unseen in other places (Teo et al. 2023).

From a perspective of state entrepreneurialism, we suggest a dialectic relation between the state and entrepreneurialism. Hence, stressing the state's role does not create such a problem (Wu 2023c). The minimalist state is only implied in the governance literature. In other words, 'governance' does not have to take an 'entrepreneurial'—meaning pro-market—form. Instead, it might be an art of governing the market. In this book, we investigate 'governance' without assuming such a connotation of deregulation. Such a tendency exists in the dynamic adjustment of regulation-deregulation, conceptualized broadly in the research on rescaled state spaces (Brenner 2019). As will be found throughout the book, the market tool and associated social mobilization have been used.

The entrepreneurial shift—neoliberalism

The concept of governance originated from the changing political-economic relation between the state, the market, and society. As commonly understood, the government is supposed to govern the market and society. However, there has been a tendency to reduce government service delivery, that is, an overall change 'from government to governance' involving collaboration between the state, the market or business, and the civil society or communities. Governance involves multiple actors: the state, the market, and society. Therefore, governance includes a number of innovations in how the government governs. There are two significant changes in government innovation: first, neoliberalization, indicating the rising role of the market in economic and social regulation; and second, the post-political society, suggesting governing through technical rationality rather than democratic political processes.

First, neoliberalism is an ideology that suggests the free market is the most efficient way of governance. The adoption of neoliberalism advocates reducing government intervention—a process of neoliberalization. According to this theory of neoliberalism, the state should retreat from welfare redistribution in order to enhance economic efficiency (Harvey 2005).

Second, the implication of neoliberalization for society is the rising 'post-political society,' away from the political purpose of the welfare state to addressing the social demand for redistribution (Swyngedouw et al. 2002; Swyngedouw 2009).[3] According to the post-political theory, the public is excluded by the process of regulation based on market principles. Consequently, the governance is characterized by exclusionary politics, the deficit of public participation, and the lack of democracy in decision-making. Large-scale urban development projects illustrate the neoliberalization shift (Swyngedouw et al. 2002). Because of their large scale and exceptionality, these projects reveal the market's rising role, especially that of property developers.

The earliest observation of this shift of governance was proposed as urban entrepreneurialism by Harvey (1989). He specifies the precise move from managerialism to urban entrepreneurialism in the capitalist society. Faced with capital mobility, cities compete with each other. Capital mobility gives greater power to investors. Cities emphasize less welfare provision, but more wealth generation and economic development. Later, in an influential book, Harvey (2005) characterizes this as increasing neoliberalism. In other words, he further specifies this as a *political* strategy promoted through deliberate state action—following the ideology to achieve its hidden class interest—to restore the profit rate and class dominance.[4]

Interrogating urban entrepreneurialism in Hong Kong, Jessop and Sum (2000) amended it with a regulation school explanation and the advancing of a 'state-theoretical' perspective (Jessop 2002b). The entrepreneurial city embodies a set of statecraft—the art of governing and staying in office. Strictly speaking, their study of Hong Kong should not be regarded as an application of Harvey's urban entrepreneurialism, as the toolkit in this 'neoliberal' city with a conventional image of 'positive non-intervention' includes a wide set of policy actions that go beyond deregulation. They describe the actual role of the city-state. The entrepreneurial city confirms three criteria: developing 'innovative strategies,' 'pursuing these strategies in a reflexive and entrepreneurial fashion,' and presenting 'entrepreneurial discourses.' These are soft, or even proactive actions. They reflect state-centered or state-engineered actions, referred to as 'inter-scalar strategies' or so-called 'glurbanization'—embedding globalization within local urban development.

There are also criticisms about the theory of urban entrepreneurialism (Ong 2007; Le Galès 2016; Pinson and Journel 2017). Some critiques suggest urban entrepreneurialism does not generally reflect state retreat but rather a particular form of urban governance. As seen in the entrepreneurial city, the state adopts specific strategies, imposes market disciplines, and uses market guidance at a distance. So-called urban entrepreneurialism does not pay sufficient attention to the state. These criticisms seem to suggest a structuralist problem of Harvey's conceptualization. However, Harvey does not believe his explanation to be incompatible with 'governmentality.' Foucauldian scholars use the term governmentality to refer to neoliberal governmental techniques, especially governing at a distance rather than through direct control. Governance is achieved through a 'mentality' of self-regulation and self-governance, which diminishes the state's role.

Similarly, rather than thinking of 'flexible accumulation' as an economic process associated with globalization, Harvey (2007) believes it to be a constructed order. He used the term 'flexible accumulation' in the sense of 'strategy' in contrast to a process of 'economic restructuring' before he arrives at the term 'neoliberalism.'[5] In this way, his interpretation seems less structuralist than is commonly believed. He almost understands neoliberalism as statecraft, as he suggests that the essential feature of neoliberalism is not about the operations of the free market but rather to achieve 'class restoration.' He points out the overall intention of neoliberalism; that is, its political dimension. The purpose is to restore profitability in capital accumulation and maintain class dominance. From his reaction to the interpretation of governmentality, we can see that he does not find that they are incompatible.[6] This position also applies to China. Market reform is thus a state strategy for survival following the political turmoil of the Cultural Revolution (Wu 2010).

According to this explanation, which has a greater focus on governmentality, Harvey (2007, pp. 28–29) suggests, 'we can, therefore, examine the history of neoliberalism either as a utopian capitalism or as a political scheme aimed at reestablishing the conditions for capital accumulation and the restoration of class power. … I shall argue that the last of these objectives has dominated.' He even suggests an extra-economic measure—'accumulation by dispossession' (Harvey 2003).

Interestingly, Harvey's interpretation moves closer to what we have seen in China by distinguishing intentionality (class restoration) and action (greater use of the market—'flexible accumulation' or 'accumulation by dispossession'). Here, we distinguish intentionality from actions, which resonates with Harvey's work. This distinction is particularly important within the Chinese context as China is still an emerging market economy without a matured capitalist system (Wu et al. 2024). For Harvey's research on neoliberalism, class restoration is about the capitalist class as the ruling class in capitalism. Capital accumulation is naturally a political project to restore the dominance of the capitalist class. However, in state entrepreneurialism, using the market does not necessarily serve the market in terms of intentionality, while the economic operation is still subject to capital accumulation. For this reason, Harvey does not elevate the state as a proactive actor (hence neoliberalism). This particular approach—neoliberalism—better serves the capitalist class interest. In his view, the state has very limited 'proactiveness' or discretion, though he later indicates accumulation by dispossession might be a forceful tactic.[7]

In Harvey's interpretation of advanced capitalism, the ultimate goal is to restore profitability and class dominance and use the market (flexible accumulation) as an instrument. In our case of governing the Chinese city, while a similar but variegated marketization approach has been launched, the centrality is state legitimacy through its role in maintaining structural coherence. However, other studies about Harvey's urban entrepreneurialism have ignored this focus on intentional politics. However, this aspect is made more explicit in Jessop and Sum's (2000) re-interpretation of the 'entrepreneurial city' as entrepreneurial discourses, narratives, and self-images. The entrepreneurial discourse narrates their cities and markets as entrepreneurial (p. 2292). More attention has been paid to the entrepreneurial side of neoliberalism. To a lesser extent, neoliberalism is understood as a political strategy. This book helps to highlight this neglected aspect of the phenomenon.

More shifts beyond urban entrepreneurialism

After Harvey's governance shift to urban entrepreneurialism, Jamie Peck (2012, 2017a, 2017b) indicates another shift: from a more entrepreneurial public sector to the dominance of finance. Urban governance is now subject to a pervasive capital logic, and he uses the term 'financialized governance.'

A related but alternative interpretation of the recent shift as 'municipal statecraft' suggests an opposite direction (Lauermann 2018). This focuses on a greater role for the state, especially at the municipal level. Hence, similar to neoliberalism's intentionality, municipal statecraft pursues 'multiple political logics in parallel with growth, in ways that are not only speculative in an economic sense but more broadly experimental' (p. 212).

Even further, we have seen a different trend of municipal statecraft with the foundational economy and operative development agencies. Conceptualized as 'municipal radicalism' (Thompson 2023; Roth et al. 2023), the notion confirms the role of the state in economic development beyond delivering public services. Still, its involvement is not driven by or limited to a capital logic. In this section, these directions of change in governance are explained.

Austerity urbanism and late entrepreneurialism

Urban entrepreneurialism has ended since the onset of the global financial crisis in 2008. Western cities' public finances have experienced budget cuts and limited fiscal resources for service delivery. Describing the situation in Atlantic City, USA, for example, Jamie Peck (2017b) reaches a mid-level theory that explains the shift 'from the entrepreneurial city to austerity urbanism and financialized urban governance' (p. 327). This transition is related to his earlier overall conceptualization of 'austerity urbanism' (Peck 2012).

Atlantic City had a long, notorious history of emulating the model of entrepreneurial urbanism. However, now it has ended up with a 'distinctly "late-entrepreneurial" moment of fiscally mandated governance and political crisis' (p. 327). Because of austerity, the state, and particularly the local state, lost its financial control. Conceptualizing through another study of Detroit, the entrepreneurial city is understood as ineffective because it has been captured by the financial logic of the capital market:

> Compounding a shift towards entrepreneurial urban governance, cities now find themselves in an operating environment that has been constitutively financialized. Bondholder-value disciplines have become systematic in reach, along with an amplified role for financial

> gatekeepers like credit rating agencies; technocratic forms of financial management have been spreading and deepening.
>
> (Peck and Whiteside 2016 p. 235)

They describe how external investors and the financial sector now control the city of Detroit. In this sense, urban governance has been *financialized.*

Fiscal austerity triggers or constitutes the basis of new urban governance, which is less performed by the entrepreneurial public sector. Urban governance is subject to a more stringent financial logic. This process is known as financialization or financializing the city, which will be discussed further in Chapter 3. With regard to Detroit, it is illuminating to see the emergent financialized urban governance.

The US case might be an outlier. Aalbers (2023) suggests various types of relations between the state and finance. For example, the governance of infrastructure investment in the UK through programs such as the City Deals does not amount to financialized governance. However, statecraft in the country has been financialized, especially at the city level (Pike et al. 2019; Pike 2023). According to this conceptualization, the new governance represents a new type of managerialism utilizing the financial market to raise capital.[8]

While global capital dominates urban development finance, local actors are still crucial in financialization. Global and local actors may work together to form an assemblage or 'trans-local financial network' (Halbert and Rouanet 2014). They interact in a relational term. Global investors rely on local actors to translate local financial practices into global standards in order to understand the associated risks. This trans-local view highlights the importance of territorial politics from a comparative perspective (Robinson 2022).

Such a trend emerges across different national contexts. This is relevant to the case of China. We see the increasing tension between financial logic and territorial politics. For China, the state has deployed financial tools, especially the land-based financial approach, in order to mobilize capital for fiscal expansion (Feng et al. 2022a; Wu 2023a). In the post-pandemic period, the Chinese economy suddenly experienced an unexpectedly slow recovery in 2023, with the result that the crisis of local government debt has been exacerbated. While we saw previously that the state has used financial logic for its own purposes, will it now be subject to a more brutal financial discipline? This remains to be seen.

Municipal statecraft and radicalism

Instead of considering urban governance following the capital logic, the state presents its own territorial logic. One view of more state-centered governance is on entrepreneurial municipalities as 'municipal statecraft' (Lauermann 2018). The municipal government agenda does not exclusively aim to promote capital-driven growth. It may have other agendas in parallel. Governance may be speculative in terms of attracting investors and experimental in different policies. These policy experiments are beyond a profit-generation purpose.

While urban entrepreneurialism emphasizes inter-city competition, municipal statecraft seeks inter-urban diplomacy and launches negotiations with other localities. The governance rationale is not dictated by growth and profit-making agenda, but includes a wide range of objectives. Municipal statecraft distinguishes between speculative and experimental approaches, paying more attention to diverse urban policy experiments. The purpose of development, thus, is not speculation on land values as described by the growth machine.

Applying this perspective to the city region, municipal statecraft explains the collaboration between different localities to form a system of city-regionalism.

The 'new' urban governance argues that urban entrepreneurialism no longer focuses on reducing regulation constraints to attract inward investment. There are a variety of forms, including urban diplomacy, urban intrapreneurialism, and urban speculation (Phelps and Miao 2020). Urban diplomacy aims to enhance a city's status by supporting entrepreneurial activities, such as marketing the city to outsiders. Unlike entrepreneurialism, which looks for external investors, urban intrapreneurialism aims to reform government efficiency, enhance coordination between government departments, and develop the economy through state-owned enterprises. Originating from observing the Singapore government, the model shows how well-resourced bureaucracies propel innovation within the sizeable state sector (Phelps and Miao 2020; Miao et al. 2023). In addition to attracting investors, the city may invest in large infrastructure projects through a process of 'urban speculation.'

In short, the new urban governance is state-centered rather than being reactionary and reliant on external investors. While the growth machine thesis explains governance through a capital-driven agenda for increases in land value, the motivation might come from the state actors and the public sector. Rather than being external to the state, the variety of urban entrepreneurialism has its own centrality, reflecting the state's intentionality.

Earlier, we suggested that fiscal austerity led to the end of urban entrepreneurialism because the local state had limited resources to deploy for economic growth. Financialized entrepreneurialism is interpreted as entrepreneurial activities originating from financial operations. This indicates the rising dominance of financial capital in urban governance. However, there is another way of thinking of financialized governance: that the state solicits finance. Or that financial instruments are created or utilized by the state through statecraft (Pike et al. 2019). This defines the nature of 'financialization' as financialized statecraft (see Chapter 4). This means that the state resorts to a financial approach. According to this explanation, the role of the state remains important. The city-state is critical in extracting values.

Municipal statecraft is regarded as a 'variety of urban entrepreneurialism'. That is, the statecraft executed by the municipality. This is different from urban entrepreneurialism in that the latter term emphasizes neoliberalism—that is, the rising power of the market. In such a system the municipalities take an entrepreneurial role. This notion is closer to the term 'state entrepreneurialism' as conceptualized in China, since the entrepreneurial methods are used within and by the state to achieve its discretionary purposes—some are more political or ecological, such as creating a 'harmonious' relationship between people and nature. In other words, they are entrepreneurial municipalities for a purpose beyond finance.

Further, beyond the varieties of urban entrepreneurialism (Phelps and Miao 2020) and other forms, such as 'financialized municipal entrepreneurialism' and 'city statecraft' (Beswick and Penny 2018; Pike et al. 2019), and 'municipal statecraft' for an entrepreneurial purpose (Lauermann 2018), a more radical turn is a societal movement toward commodification forms under 'municipal radicalism,' taking cooperative and functional economies (Thompson 2023). This municipal-cooperative form of development has its own territorial logic beyond the logic of external capital accumulation through the relationship between local government and civil society. This form goes beyond state-centered governance, an entrepreneurial public sector under new public management. This confirms the end of urban entrepreneurialism, but disagrees with the dominance of finance or the market. Instead, it describes the role of civil society.

Temporal and geographical varieties

Neoliberalization should not be regarded as a linear, homogenous, and universal global trend. The ascendancy of market roles has experienced different phases and presents diverse geographies. There are both temporal and geographical variations. Regarding the role of the state in this process, it is possible to detect a distinct phase transition from earlier roll-back to later roll-out neoliberalization (Peck and Tickell 2002). The phase of rolling back means reducing state intervention, while the rolling-out phase involves the state actively adopting more measures to support market development. In this later rolling-out phase, the state actually plays a stronger role in enforcing market rules and extends the scope over which market provision is applied. Thus, the advancement of market rationality is not incompatible with a strong state under neoliberalism.

The process of neoliberalization shows a feature of path dependence and geographical variations. In the Global South, there is a tradition of greater informalities. The governance of urban development involves wider formal and informal relations. However, informalities in governance are not solely a feature of the Global South. It is argued that increasing informality is more pervasive, reflecting a mode of global neoliberalization (Roy 2005).

In more advanced capitalist societies, neoliberalization generates new exceptions to rules and a lack of democratic accountability. In capitalist societies, civil society has been a vital force. In the Global South, urban politics do not adhere to formal procedures and middle-class civil society. For example, in India, the role of communities is represented through the notion of 'political society' associated with electoral politics (Chatterjee 2004). Instead of middle-class-based politics, different social groups are mobilized in everyday life, while dealing with the urban environment and struggling politically.[9] In Western democratic societies, by contrast, neoliberalization encounters social activism, insurgent citizenship, and variegated forms of community engagement. The interactions are influenced by a political agenda, which might not be neoliberalism (Hilbrandt 2017). Social contestation happens similarly in China, but no social movement has been formed.

From the studies conducted in the Global South, the conceptualization of neoliberalism is criticized. For example, when examining urban development policies, it is difficult to categorize them into neoliberalism toolkits (Parnell and Robinson 2012; Parnell and Pieterse 2016). Some suggest that, for example, Indian states act like East Asian developmental states (Doshi 2019). During the implementation of development policies, local communities and non-government organizations (NGOs) both play important roles. The literature on the 'right to the city' emphasizes social agencies of everyday life (Weinstein and Ren 2009).

Looking at urban development in South Africa, for example, urban politics can be characterized as 'private urbanism' (Murray 2017), as shown in the development of the Waterfall City—a super-large estate in the periphery of the metropolitan area of Johannesburg (Murray 2015). The project has been privately built into a 'gated community.' This is an exemplar of neoliberalism in community governance. That is, neoliberalism is achieved through the materiality of residential development.[10] This place reveals private urbanism because services are provided through the market rather than the municipal government; that is, they are 'privately provided' rather than being common or public goods.

However, in the same city of Johannesburg, different examples show the role of the state, or, more precisely, the activities of politicians with their own agenda. In Johannesburg, a mega-urban project known as the 'Corridor of Freedom' represents a 'people-centered city' vision (Harrison et al. 2019; Robinson et al. 2021). The project aims to use transit-oriented development (TOD) to link existing communities in the metropolitan region.

More importantly, the project was proposed by then-Mayor Moho Parks Tau (who was in office between 2011 and 2016). This differs significantly from mega-projects such as Canary Wharf (see the later discussions on mega-urban projects in Chapter 2 and redevelopment in Chapter 5).

In contrast to the image of glittering office towers in global cities, the 'marketing' materials are a series of cartoons, showing communities which are safer and connected rather than isolated. In all these cartoons, ordinary black people participate in bustling community activities on the streets. The concept of TODs, which originated in the US, was borrowed and adapted rather than for the purpose of real-estate development. Although the project also encouraged some real-estate development, like student housing along the transport route and stations, the project was designed to be funded mainly through municipal tax as Johannesburg operates a single-tier tax system (Robinson et al. 2021). The project remained incomplete because the Mayor lost his election. This shows that the project was less driven by real estate dynamics or the 'city for profit' (Shatkin 2017). Rather, it was aimed at a political achievement, even an 'achievement in office' by key politicians, which does not sound dissimilar to the Chinese notion of 'planning for growth' (Wu 2015b).

Another example is the mega-urban project in India (Kennedy 2015; Kennedy and Sood 2016; Sood and Kennedy 2020). The Indian economy has been undergoing liberalization since 1991. This has seen special economic zones being set up. Many Indian cities wish to become 'world-class' cities through learning lessons from the West or Shanghai's Pudong District or Shenzhen Special Economic Zone in China. Launching mega-urban projects can be seen as a 'state strategy for urban transformation' (Goldman 2021). These new policies aimed to create new economic spaces in the form of technopoles, cyberports, and special economic zones. In these spaces state sovereignty is excepted (Ong 2006). Many mega-urban projects are indeed for-profit and generate benefits for politicians (Shatkin 2017). In many ways, such large-scale urban developments fit the storyline of neoliberalism in the similar way they attract investors. However, the development process is more complicated. An understanding of the process reveals the underlying logic.

Urban transformation triggers a new type of urban politics. The market reform and development led to the mobilization of social groups. Indian society has a complicated structure, being divided across castes and religious, ethnic, and linguistic groups (Chatterjee 2009). Rather than an emergent civil society, it is characterized by a 'political society' (Chatterjee 2004). For example, the poor who live in slums are mobilized because local politicians exempted them from the rules in exchange for ballot papers. There is a great deal of informality in urban governance, not just inside the slums but also within the new middle-class residential projects because of wide exemptions and irregularities. The informality is described as 'occupancy urbanism' (Benjamin 2008).

The informality and occupancy urbanism is an alternative narrative to neoliberalism because the poor in India can get protection outside market forms. The extra-legal claims for land create difficulties for the state and corporations. Land acquisition becomes a conflictual process. For example, the Tata Group, the Indian car manufacturer, invested in a factory in Kolkata but it failed to clear the site because of farmers' protests (Roy 2011). When viewing the actual politics on the ground, it may not be precise to characterize such messy politics as neoliberalization. In general, the Indian context shows a strong societal role. The legacy of political fractions inherited from the colonial era still affects urban development in this post-colonial society in India.

The literature on urban governance in East Asia highlights a relatively distinctive feature of the developmental state. The developmental state actively uses industrial policies to forge the collaboration between the state, large enterprises, and banks. Observing the powerful Japanese department of the Ministry of International Trade and Industry, bank–industry collaboration under state industrial policy for priority sectors is believed to be a model that is behind rapid economic growth and urbanization. The development state also secures social conditions for economic growth by providing public housing, promising lifelong job stability, and greater workplace coherence.

Recent studies reveal that East Asian economies also experienced neoliberalization and turned to the promotion of neoliberalism (Park et al. 2012; Doucette and Park 2018; Chen and Shin 2019). But this is not conclusive. For example, the financial sector has become increasingly financialized. The state also reduces levels of investment in public housing, and the role of state-centered urban and regional policy declines. Yet the urban and regional policies still differ significantly from the Western market economies. East Asian economies have adjusted their development approaches in a post-developmental era. Fundamentally, the developmental mentality persists, perhaps reflecting a new 'state capitalism.' Neoliberalism might still be a valuable perspective for understanding the dynamics of development, as adjustment is the state's reaction toward the crisis of the developmental state. For example, balanced regional policies are no longer pursued. The global financial crisis forced East Asian economies into a neoliberal trajectory. However, the global financial crisis triggered a strong state reaction through the introduction of the stimulus package in China.

However, other scholars in economic geography doubt that neoliberalism is a useful and adequate explanation of uneven capitalist urban development and economic changes (Yeung 2019b). The economic structure change is linked to the global production network (GPN) within which their economies are embedded. The state could no longer play a decisive role. Rather, the larger and leading firms embedded in the GPN have much greater impact. This view confirms that the capitalist state serves economic development interests as a market economy. East Asian economies have a greater integration with the global market. Because of GPN and cross-national ownership, it is argued that the developmental state was hardly able to formulate an effective industrial policy to support 'domestic' enterprises because they evolved into transnational corporations. In this post-developmental era, it is difficult to formulate an effective industrial policy for the national and domestic industries.

In addition to industrial policies, the state also intervenes in land supply and has interests in land development (Hsing 2010). It is termed the 'property state' because the state is interested in land value appreciation (Haila 1999, 2016). In addition to industrial policies, research attention is paid to urban policies related to land development. The property state uses its land policies to guide urban development. The policies focus on industrial and regional development and the urban studies literature began to pay attention to property development (Shatkin 2016, 2017). This is a salient driving force in overall economic growth in East Asia. It has become so central to capital accumulation that some scholars argue that the developmental state as a public policy institution is a property regime of accumulation (Haila 2016). In other words, the state is not an industrial developmental state but rather a property developmental state with a real estate interest representing a different mode of city-building:

> Real estate has an important role in the function of the whole economy. Transnational property and development companies are important players in Singapore and Hong Kong. They form an important part of the local stock market and have enjoyed considerable growth while providing substantial revenue for governments and wealth for individuals.
>
> (Haila 2000, p. 2241)

The perspective of the property state seems to suggest that the state is embedded in land value (Hsing 2010). The motivation for development comes from this interest in land. Instead of pursuing an overall economic growth objective, the state becomes less of a public policy maker and seeks its interests in land development. Mainland China has demonstrated a similar model of property-driven development.

Hong Kong is such a case. As a high-density city, it is perceived as a so-called 'property regime' (Haila 2000, 2016) because real estate plays an important role in the economy. Land leasing is a source of public finance, and the land ownership system is unique. The government controls the ownership, but the land is leased within a restricted amount to developers in a competitive market.[11] In Hong Kong, the real estate sector has become an integral element in the 'regime of accumulation' (Smart and Lee 2003). The Hong Kong economy is driven by finance and real estate, and state intervention follows a capitalist logic (Hofman and Aalbers 2019). The real estate sector provides a high proportion of public finance and is also a major economic player, influencing other sectors. The property sector is linked to the stock market because major development companies are listed there. Because of the constraints over land supply, land development is still subject to state regulation, and land maintains its high value in this densely populated city.

Similarly, the Singaporean state maintains its influence over development through the development of real estate policies (Haila 2000). The city-state has a distinctive feature of public land ownership and high rates of public housing. Public housing is a mainstream form of housing in the overall system of housing provision. The state maintains a stable real estate market while building a globally competitive city. The state actively uses the property sector to regulate developments through a range of real estate policies (Haila 2016). One of the most recent examples of mega-urban projects in the city was the development of Marina Bay Sands.

The East Asian cases present diverse models of urban development. Compared with Europe and the US models, Hong Kong and Singapore present different models in terms of both development strategy and landownership (Haila 1999). In the US, a strong private property right creates a civic boosterism of urban governance. Europe has a stronger role in town planning and regulating the real estate sector with a similar strong private property model. However, both Hong Kong and Singapore have areas of public land ownership. The ownership itself becomes a regulatory mechanism.

Regarding the development strategy or the accumulation type, land contributes to public finance as a budgetary mechanism in Hong Kong. By contrast, Singapore has more direct regulatory approaches through real estate policies. These models highlight a variegated form of neoliberalism, either through the deviation from private land ownership or via direct regulatory control over land development. In a way, the typology of the urban development regime shows that the US growth machine with stronger private property ownership and entrepreneurial governance is a fairly specific case. The model of urban governance across the world shows a great deal of variety. The variegated land market and policies suggest no standard urban development and governance model. China has seen prevalent property interests, but also a visible state.

Because real estate occupies such an important position in economic growth and its development generates local government revenue, East Asian urban politics are centered upon property development. As such, there is a real estate aspect to policy in Asia—cities are made for profit (Shatkin 2017). Within this, mega-urban projects are used as an essential instrument. The government adopts a land management strategy to extract value from the land. In the cities for profit, land and property development are state-led but increasingly profit-driven.

In other words, the center of urban governance has shifted to a profit-driven approach. The description fits well in Chinese urban development—'land-based finance' (Hsing 2010).

One example in Shatkin's case studies is Chongqing, a city under the central government's direct jurisdiction in western China. This case reflects a 'state capitalist growth machine.' However, Shatkin (2017) also points out, 'as Chongqing's case illustrates, land commodification not only becomes the central objective of urban politics, but it also becomes a tool that the state applies to a range of social and ecological problems' (p. 210) and 'in general, China's model of urban development grants great powers to the state to realize a variety of goals' (p. 211).

In other words, individual mega-urban projects for land-driven development following the capital logic aim to generate profits. Nevertheless, the state-engineered program is not simply for profit but also intended to achieve social and political objectives. This interpretation is slightly different from the post-political perspective, which suggests that traditional democratic politics are reduced, and the political dimension has disappeared. This follows the definition of 'state capitalism' as a system 'in which the state functions as the leading economic actor and uses markets primarily for political gains' (Bremmer 2010, p. 5), Shatkin (2017) suggests that the Chinese state demonstrates a system of state capitalism and that it is used for land value extraction.

The research into East Asian property states reveals a political dimension. Similarly, in a very different context in the UK, under pressure to provide affordable housing, local councils in London set up their own development companies to manage services (Beswick and Penny 2018). The entrepreneurial municipality extracts the land value to achieve other objectives (Robinson and Attuyer 2021). Shatkin (2017) stresses the state's role in extracting land value, but shows that its purpose is not confined to capital logic. It also seems to encompass a political agenda. It is important to pursue this line of thinking in understanding urban governance. According to the perspective of state entrepreneurialism, land-profit making is not an ultimate development goal (Wu et al. 2022b). Promoting land development also has a political dimension. As seen in the case of Chinese city planning, land development differs from the North American growth machine. The dialectic interaction between the state and entrepreneurialism is a key feature of state entrepreneurialism. While all economies are influenced by this dialectic interaction between state and entrepreneurialism representing the force of capital accumulation, the exact combination form is all specific or unique, as shown by China's model of urban governance.

Summary of the perspectives

Urban entrepreneurialism is a 'new urban politics' in state capitalism. This emphasizes competitiveness. The notion of the 'entrepreneurial city' suggests that cities, like firms, compete with one another for mobile capital. Recently, however, this public-sector entrepreneurialism is claimed to have ended because of the rising dominance of financial capital. However, financialized governance under austerity urbanism does not entirely overturn entrepreneurialism. Institutional austerity puts greater pressure on local entrepreneurialism, as seen in the UK, leading to financialized entrepreneurialism—playing the game according to financial rules (Peck 2017b). According to this explanation of rising capital, urban governance has been financialized, indicating a shift from entrepreneurial to financialized governance. This interpretation is based on continuing market dominance, or the capital logic.

Further, urban entrepreneurialism has been under pressure because of rising social movements, particularly at the urban scale, such as municipalism or municipal statecraft (Thompson et al. 2020; Thompson 2023). This view emphasizes the territorial logic, i.e., the role of the state in response to the challenge to capitalism and society.

These views help to re-interpret the political dimension of urban governance. The state adopts various entrepreneurial strategies to support capital to extract land value from urban development, known as the 'city for profit' (Shatkin 2017) or the urbanization of the local state (Hsing 2010). Nevertheless, the city for profit implies the capital logic. This might not accurately describe existing urban politics, as it exaggerates the motivation for profit. In the postcolonial society of the Global South, the political dimension, the politics arising from 'communities,' is salient. Postcolonial studies present a different narrative from the post-political description of Western economies, where politics are subject to formal and technical processes. This raises a question about the purpose of extracting value, which can be political beyond a capital-based interest for capital accumulation. In other words, it is a political question. Here, the literature of critical urban studies on the shift of governance after urban entrepreneurialism is illuminating because, for scholars on China in particular and on planning studies in general, a political question helps to reveal the state's role in governance, of which the sphere of planning is technical but also political. It is a political question approached from three theoretical stances: class restoration, radical municipalism, and everyday politics. Here, China resembles Harvey's class restoration—to maintain the state's interests.

Perspectives on changing urban governance in China

Previous sections introduced the concept of governance and its variegated forms across different histories and geographies. This section focuses on changing urban governance in China, considering various concepts and explanations which have been applied. The purpose is to demonstrate a wide spectrum of explanations and their positions to situate state entrepreneurialism within this body of literature. This also helps to define some key terms of usage, such as entrepreneurialism. The term can reflect business innovation in profit-making and public administration experiments as 'new public management.' As discussed earlier, state entrepreneurialism describes an intentionality beyond profit-making. Hence, entrepreneurialism here is not limited solely to market behavior, although Chinese state entrepreneurialism is characterized by greater market tolerance and deployment.

Entrepreneurialism refers to both private and public sector innovations. According to this definition, it often refers to the market behavior of private entrepreneurs and administrative endeavors to gain a market advantage. The latter may involve non-market means or an end beyond capital accumulation. It excludes welfare or philanthropy.

The earlier studies in the 1980s and 1990s noted market re-orientation, fiscal decentralization, and, consequently, the entrepreneurial behavior of the local state.[12] The 'local state corporatism' thesis describes rural cadres' engagement in the operation of township and village enterprises (TVEs) (Oi 1992). They played a dual role of being both state officials and business people, being people who have collectively transformed the local state into an entrepreneurial entity. Even a more structural view suggests the institutional source of entrepreneurialism—the fiscal responsibility of local government, together with greater local decision-making autonomy, turned local governments into entities like entrepreneurial industrial firms (Walder 1995). The market reform introduced business opportunities to city officials. Establishing a real estate market in urban areas led to the local state's exploitation of market opportunities (Duckett 2001). These explanations for emergent entrepreneurial behavior are similar to a structural explanation of 'land finance' to be explained later in this section—continuing on the fiscal reason and economic behaviors of the local state.

In order to provide a comprehensive review of the specific Chinese literature, this section includes an ideological explanation (neoliberalism), a political explanation (the 'GDP tournament'), an economic explanation ('land finance'), a management explanation (the 'shareholder state' and financialization), a state theoretical explanation ('market in state'), and various governmentality explanations (fragmented authoritarianism and bargained authoritarianism).

Neoliberalism

Neoliberalism is an overarching explanation for China's market reorientation after economic reform. This explanation includes a wide range of popular descriptions about adopting the 'Washington Consensus' and the debate over its applicability within the Chinese context (Wu 2010). The key text here is David Harvey's (2005) influential book *A Brief History of Neoliberalism*, which describes China's governance as 'neoliberalism with Chinese characteristics.'

Economic reform within the country has created a process quite similar to neoliberalization, which is believed to capture the fundamental features of China's urban governance. Nevertheless, along with market-oriented experiments, China also maintains some authoritarian features, hence the notion of uniquely 'Chinese characteristics.' This raises the question of whether the Chinese characteristic is a more fundamental feature than neoliberalism, namely a governmentality perspective pointing to the use of market reform as a governance technique (Ong 2006). It is thus a state authoritarianism with neoliberal characteristics. As noted with regard to state entrepreneurialism, this switch of denomination and denominator moves closer to entrepreneurial statecraft.

Earlier Chinese urban studies presented a wide range of housing and land market practices. Along with market reform, urban redevelopment is regarded as the 'frontier of neoliberalism' (He and Wu 2009), triggering the demolition of old neighborhoods, the displacement of existing residents, growing social inequality, and residential segregation.

However, there are critiques about using neoliberalism as a theory to explain China's urban development. Trying to modify the neoliberalism explanation, this body of literature argues that we must pay attention to the variegation, as China's neoliberalism is 'complex and heterogeneous' (Peck and Zhang 2013). Instead of attempting to classify China's model of urban governance as a template of the nation-state as the variety of capitalism (VoC) (Hall and Soskice 2001), it is more appropriate to understand its variegated and combinational feature (Peck 2021), which may contain regional sub-models and variations (Zhang and Peck 2016).

In addition to this modification through 'variegated neoliberalism,' there are doubts that neoliberalism over whether it could be regarded as a sufficient explanation (for example, Keith et al. 2014; Buckingham 2017; Zhou et al. 2019; Phelps and Miao 2020). Some examples in digital technology, green innovation, and grassroots entrepreneurs point to a 'spontaneous recombination and reconfiguration of the urbanizing society at the grassroots level' (Zhou et al. 2019, p. 39) and the role of the state and society.

GDP tournament

In addition to neoliberalism, regarding the emergence of entrepreneurial governance, there are two specific explanations from economic perspectives. The first concerns the behavior of state cadres who act rationally to maximize their career prospects. This 'behavioral' explanation stresses the motivation of state officials or political leaders, namely their career promotion.

In the process of inter-city competition for economic growth, known as the 'GDP tournament,' these officials become entrepreneurial and adopt governance innovation (Li and Zhou 2005). Critical yet empirically challenging evidence is that the target of GDP growth, as an 'achievement in office,' is used as a major criterion for cadre performance evaluation.

Although this explanation emphasizes economic performance, it is actually a political explanation. It goes beyond machine politics in which the government supports the private sector to maximize land values. The competitive environment and cadre evaluation are all engineered by the state. However, whether or not this hypothesis confirms the politics in reality is still questionable. There is a vast literature to test this hypothesis and controversies about conceptual limitation and empirical invalidation (Su and Tao 2020). Economic achievement is only one aspect of cadre evaluation. Other important factors for promotion may include the alignment with central government policies, or loyalty to a particular political faction. Thus, the focus on GDP-ism would appear to be too narrow. The explanation seems to capture some features of the particular political and economic environment during the 'growth-first' era of Deng Xiaoping.

Land finance

Land finance is another widely circulated explanation for entrepreneurial governance. This explanation is similar to the thesis advanced by the 'growth machine,' stressing the dynamics of land revenue generation to meet the demand of local fiscal expenditure because of fiscal disparity and the gap between local revenue and expenditure (Tao et al. 2010; Su and Tao 2017). As a structural explanation, this refers to particular Chinese institutions that explain the emergence of entrepreneurial behavior within local government. These include the tax-sharing system, the local state's reliance on taxation rather than finance as it cannot raise capital directly from the financial market, and the state's control over land supply to create a monopolistic rent.

Driven by fiscal incentives, local state officials demonstrate entrepreneurial behavior. They are motivated by economic considerations over land income. The underlying dynamic is revenue maximization—maximizing the revenue from the land. This explanation seems to be supported by the land revenue data and the observation that the local state has become 'territorialized' or urbanized. The 'urbanization of local states' suggests that the state is interested in capturing land value through urbanization (Hsing 2010).

However, the actual operation of government may be more complicated. For example, the motivation may not be confined to short-term revenue generation. The state has more strategic incentives than land revenue maximization due to the Chinese state's multi-scalar nature (Zhang et al. 2022; Zhang and Wu 2022b). Whether the Chinese local state conforms to a growth machine is questionable, however. As will be seen later in Chapter 3, the local state demonstrates an entrepreneurial statecraft — 'planning for growth.' The notion of 'growth' here is beyond profit-making, but remains a centrally-guided economic development goal.

To maintain state power, the local state has to align with central government policies and become part of the state apparatus. Within the country, there are complex interactions between different levels of government. The central government maintains its power by endorsing special policies, permitting exemptions, appointing local leaders, and exerting hierarchical control. In the meantime, the local state maintains discretion through state-sanctioned informality and informal practices excepted from central control (Wu 2018b, p. 1385).

Land value maximization may be a very important aspect of the operation of the local state, but it is not an exclusive consideration. While the thesis of land finance seems to be quite different from the GDP tournament, they both rely on the interpretation of the local state motivation. The GDP tournament tends to focus on the careers of individual political leaders, and land finance is more structural about the collective interest of the cadre class in maintaining the local state apparatus, like a growth machine. The municipal government leaders may focus on political careers, while the grassroots state officials have to deal with the operation and tend to have greater revenue concerns.

The shareholder state

From a public management perspective, the Chinese state involves market coordination based on property rights and asset ownership. This form of state is conceptualized as the shareholder state (Wang 2015). The literature focuses on the changing state itself, especially the changing state-asset management. In the shareholder state, the management of state assets has changed to a corporatist style. With the introduction of corporate management, economic management has been financialized. Evidence of this is that state assets are under the control of the State Asset Supervision and Administration Commission (SASAC), which manages many large SOEs. This interesting perspective suggests that the state resorts to 'financial means to manage its ownership, assets and public investments' (p. 603). This conceptualization is similar to the new public management literature and its application to urban governance as 'new urban managerialism' (Phelps and Miao 2020) and using the market to reform the state (see below).

Market in state

The 'market in state' thesis explains the actual operating market in China (Zheng and Huang 2018). In the capitalist economy, it represents the state in the market. That is, the state intervenes in the market operation. In contrast, market in state is a 'system of political economy whereby a substantial part of the market and market mechanisms is firmly embedded and confined within institutional mechanisms of the state' (p. 126). According to this conceptualization, China's state activities are participated by market actors and through market approaches. This notion is very close to entrepreneurial statecraft, where the state operation resorts to market means or actors. The thesis describes how the state absorbs the market into its governance. That is, how it builds the market within the state.

According to the market in state thesis, there are three levels of operation: the grassroots level, the middle ground level, and the national level. At the grassroots, China shows a system of free market capitalism. At the top is state capitalism, and a middle ground is in between, where the nexus between the state and private enterprises is established. The monetary regime controls the overall economy at the national level. Large SOEs are subject to state control through state ownership. The operation is, according to this thesis, state capitalism. State-owned enterprises are the features of Chinese state capitalism (Naughton and Tsai 2015).

All in all, market in state describes a variegated combination of the state and market, and even the market is not self-regulating but a state-centered order. The source of state dominance comes from centrally managed SOEs that contribute to fiscal revenue, and their executives are political actors as party-controlled cadres.

Governmentality of authoritarianism

The governmentality perspective focuses on the relationship between the state and its subjects (that is, citizens in society). It stresses the nature of authoritarianism in China. This body of literature does not focus on the state's entrepreneurial behavior; rather, it describes its changing behavior in dealing with society. The literature in China studies has further moved into a more nuanced understanding of decentralization and negotiation after economic reform, known as either 'fragmented authoritarianism' or 'bargained authoritarianism.'

China is often characterized as an authoritarian state. Regarding the level of centralization, recent studies have begun to modify this notion and noted a less authoritarian pattern. For example, the notion of 'fragmented authoritarianism' explains the hierarchical and horizontal divisions inside the state (Lieberthal and Lampton 2012), and its 2.0 version indicates additional components such as the media, non-government organizations, and peripheral officials becoming 'policy entrepreneurs' (Mertha 2009). The power is not centered upon a single government branch. Hence, coordination between departments and divisions becomes necessary. As a result, it demonstrates certain features of governance.

Another influential conceptualization is 'bargained authoritarianism,' which focuses on the state–society relationship (Lee and Zhang 2013). The state attempts to maintain social stability and has to bargain with society. As a result, it has to compromise and maintain certain flexibility. The model describes China's urban governance through state–society interaction, i.e., popular protests and bargaining. This governmentality literature suggests some flexible, experimental, and participatory features of urban governance. The state is not confined to the entrepreneurial notion; it also deals with social issues.

Governance as a mode of regulation maintaining 'structural coherence'

From a political economy perspective, we can understand governance as a mode of regulation. The mode of regulation, originating from the regulation school, needs to suit the political economic system, or, more specifically, the 'regime of accumulation'—the system to extract surplus value (MacLeod 2001). The pair concepts—the regime of accumulation and mode of regulation—have been applied to analyzing the changing governance in 'transitional economies' that are under the transition toward market economies. Commonly noted in the 1990s, they are Central and Eastern European economies and the Chinese economy. The changes are parallel in the East and West (Wu 2003b) (Figure 1.1). As the figure shows, the Fordist regime in Western capitalist economies is characterized by the economy of scale and a consequent Keynesian welfare state. Large-scale production with standardization is performed by big

The socialist regime Extensive accumulation through state-led industrialization; redistributive state	**The Fordist regime** Economy of scale; Keynesian welfare state
The post-socialist regime Intensive accumulation through commodification; spatial fix (export orientation); the entrepreneurial state	**The post-Fordist regime** Economy of scope; flexible accumulation; neoliberalism; post-Keynesian workfare

Figure 1.1 Governance as a mode of regulation, changing urban processes and the shift of governance mode.

Source: Wu 2003

factories, related big workers' unions, and, consequently, a big government. The Keynesian welfare state is developed to ensure the 'structural coherence' of the political economy.

The mode of regulation changed when the accumulation approach became more flexible—transforming into a flexible accumulation regime. In the post-Fordist regime, accumulation pursues the economy of scope. Instead of producing large quantities, products become differentiated and are manufactured on a just in time basis. This regime also led to capital mobility, generating its equivalent neoliberalism ideology. The mode of regulation changed to neoliberalism and post-Keynesian workfare. This is an established explanation of governance as a mode of regulation.

Applying this perspective to post-socialist economies, we observe that the socialist regime is characterized by extensive accumulation through state-led industrialization. Developing alongside state-led industrialization was the redistributive state, which allocated resources for production and arranged social reproduction, such as providing public housing to workers. The post-socialist regime witnessed intensive accumulation through commodification and spatial fixes (for example, extending the geography of accumulation through export orientation). The production system has subsequently been reoriented through marketization to produce commodities for the global economy. The local economies are linked to the world through global production networks. The local state has become more market-friendly—moving toward the entrepreneurial state. This explains the change from an inward-looking and administratively centered mode of regulation to a more collaborative mode, particularly at the local level. This explanation of political economy focuses on transforming the economic system and related changes in governance approaches.

Applying this perspective to examine the 'world-factory' model in China, we can understand the foundation of Chinese state entrepreneurialism as a mode of governance (Wu 2017) (Figure 1.2). This is a political economy reading focusing on the development model to explain the type of governance.

The accumulation regime is characterized by a close association between local urban development and the global economy. Using the relatively lower prices of land and labor costs, Chinese cities developed industrial capacities. Through re-engineering local governance, an entrepreneurial local state ensures the supply of low-cost land to attract investors for industrial activities. In turn, the supply attracts investment and rural migrant workers. Foreign direct investment has been a major source of capital, in addition to a high rate of domestic savings. This development model is supported by a particular mode of governance which

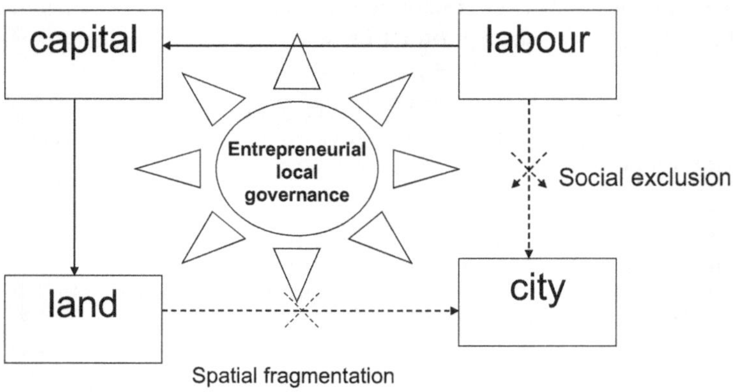

Figure 1.2 The world-factory model of China and related governance form of state entrepreneurialism.

Source: Wu 2017

requires the state's capacity for land supply. This is achieved through its monopolistic position in terms of land acquisition for sales (Su and Tao 2017).

Meanwhile, rural migrants were excluded from social provision, maintaining their status as guest workers. The social reproduction of migrant workers is entirely left to the market. Because land is released to attract investors, it creates a production city, leading to large parcels of industrial development, the excessive use of land for industrial uses, rapid urban expansion, and spatial fragmentation. The state's role is visible in managing both land and labor production factors. Through labor management, costs have been maintained at a low level, lagging behind economic development. Because of this constraint, the labor force does not constitute effective market demand. This departs from Keynesian capitalist economies' mass production and consumption because the economic development model relies on overseas markets to expand capital accumulation. This embedded contraction leads to future crises.

The model also relies heavily on investment in fixed assets and infrastructure. Infrastructure development is driven by land value capture, assuming a constant value appreciation of future land. Local governments have strong incentives to acquire rural land in order to set up various development zones to initiate urban development and further capture land value. This model may sound straightforward, but it contains a number of contradictory tendencies. The model explains that the local state becomes more market-friendly because of the linkage with the global economy. At a deeper level, however, the model reveals the state maneuver—using 'market instruments' to realize its intention rather than just being market-friendly. This is more state-centered governance. The market logic does not necessarily capture the state.

Reflecting on the world-factory governance model, we understand entrepreneurial statecraft—state entrepreneurialism. This is a political economy understanding of governance, as the Chinese economy is now embedded in global commodity production, operating as a world factory. Its governance has consequently been transformed to become more market-friendly with neoliberal tendencies.

On the other hand, the model also implies intentional strategies, that is, state entrepreneurialism. The state uses its control over land and releases the land to attract investment, which might be regarded as governing for the market. However, it is also an institutional building, intentionally using existing land institutions to achieve more strategic objectives, such as developing and upgrading the economy and achieving the Chinese dream. In other words, the world-factory model is not just a reactionary one. Like the developmental state, it reflects that the state engineers or guides development to create a new development trajectory. It is not only for development but also very often for coping with the crisis, as occurred, for example, during the Asian financial crisis in 1997 and the global financial crisis in 2008. As will be seen in Chapter 3 on financialization, the global financial crisis triggered financialization in China. The state's stimulus package as a crisis management strategy indicates that although the influence of the finance sector has also been strengthened along with financialization, the state is not simply captured by the capital logic of the financial market.

Thinking of governance as a mode of regulation, we reveal that planning centrality is a feature of state entrepreneurialism. However, this does not mean that everything is organized according to plan. Informalities prevail in governance. Looking at governance practice across spatial scales, we understand this is not a top-down process. People living in Chinese cities have their agencies. China's urban transformation has had immense impacts on people's lives. Urban dwellers and rural migrants overcome regulation constraints and create a space of their own. Residents try to develop a new life. For example, migrants settle down in the urban villages based on their origin, in places like 'Little Hubei' in Guangzhou, where people from

Hubei province live together (Liu et al. 2015). They developed a trans-local business network and operated small cloth-making and garment workshops. Linking with other people in Hubei, they chose to live in the same place. Migrants residing in enclaves maintain active socializing (Wu and Logan 2016). Their neighboring and socializing give them mutual support. Thus, they survive in a more challenging and sometimes discriminatory living environment. Thus, it is clear that planning centrality does not eliminate informality. Even when it wishes to, it could not effectively do so. Very often, planning policies aim to demolish informal settlements. The government, however, has to tolerate informal development because of pragmatic considerations. This is also a result of the mode of governance. Informality exists because the government can no longer control all means and resources as it has to resort to 'market instruments.' The state has to relax control for practical reasons.

State entrepreneurialism needs to mobilize various actors and use market resources. In this sense, governance rather than government affects the course of urban transformation. Given the decentralization of resources, the state no longer determines urban development. The state has to collaborate with other actors. Urban villages are one good example of the state's inability to provide affordable housing to rural migrants. Consequently, the private informal rental market plays just such a role. Significant informality is thus inevitable. The governance approach means that the state has to compromise and give space to people's agencies in urban development. In short, state entrepreneurialism does not mean the return to planning. Ordinary people still have their agencies which affect the outcome of urban planning. State entrepreneurialism thus includes two types of agencies. First are the agencies of political actors, as Harvey describes the capitalist class. Second, there are the agencies of ordinary people, which have been 'strategically selected' (practically, introduced, and tolerated) by the state (Jessop 2002b).

Critical reflections

This chapter introduces Harvey's insightful observation of the shift toward urban entrepreneurialism and discusses further turns afterward producing either financialized governance or municipal statecraft incorporating social movements. The chapter also briefly introduces the term 'state entrepreneurialism' to characterize the pattern of Chinese urban governance. In this respect we distinguish state entrepreneurialism from urban entrepreneurialism because of the visible role of the state in the former system. Nevertheless, the role of the state in capital accumulation and, in turn, political dominance is now a global aspect of state capitalism (Kinossian and Morgan 2023).

As introduced earlier, urban entrepreneurialism has come to an end. It has not become just variegated but also mutated into financialized governance (Peck 2017a, b). The new regime is qualitatively different because there is a sharp contrast between entrepreneurial city strategies and systematic financialization. Then, how does state entrepreneurialism in China differ from financialized governance under state capitalism?

State entrepreneurialism is similar to state capitalism because both have seen the expansion of financial instruments and financial statecraft. However, state entrepreneurialism stresses the state's rationality in introducing financial logic, rather than representing such logic. The global financial crisis has similarly transformed the system of Chinese urban governance. In response, the Chinese state launched a stimulus package — triggering the process of financialization in China. The world-factory model thus mutated into a debt-driven infrastructure model. State entrepreneurialism has not come to an end. It has been strengthened.

In the world-factory model, the local state aimed to attract investment and build cities around global industrial production. Under the stimulus package, local governments were asked to find finance themselves. This led them to borrow capital from land to initiate infrastructure-led development. Their behavior began to deviate from financial logic. The agents for this are urban development and investment corporations, originally more or less ordinary developers, who have now become financial vehicles. Through these vehicles, the state is more deeply involved in financing urban development—from creating finance to organizing development projects. Thus, deploying financial instruments imposes a new operational requirement for state entrepreneurialism. The use of local government financial vehicles reflects the feature of state entrepreneurialism, which is different from earlier more profit-oriented entrepreneurial activities of township and village officials. In short, state-owned financial organizations are agencies of state entrepreneurialism, representing some state rationality rather than market efficiency. Their reckless leverage in the financial market is the source of financial risks.

Reflecting on 'governance,' state entrepreneurialism means that the state uses the market and consequential greater tolerance toward informality and society mobilization to maintain its centrality—to achieve more intentional outcomes. This means that state entrepreneurialism can also be seen from a governmentality perspective, just like viewing neoliberalization as a governance technique. However, such a technique is conditioned by structural force. While we emphasize intentionality and discretion (because of sovereignty), state entrepreneurialism manifests the imperative of capital accumulation. China has limited choice, but opens its door to the world and strengthens the state's role in dealing with its consequences. Introducing market mechanisms and social changes under market reform generates a visible state (Wu 2022a).

State entrepreneurialism does not mean relocating power to the market or society. It is not about power redistribution. It is about using market mechanisms and societal involvement for state purposes. Sometimes, this requires reconfiguring existing institutions.

The reconfiguration of existing institutions through innovation and flexibility creates a notion of 'governance.' The concept of governance here means horizontal and vertical coordination within the state and its relation with the market and society. The state is also more than a single and unified entity. The state contains complex relations across scales as formalized social relations. For example, there are multi-level governments with sectoral and territorial administrative units across different hierarchies, forming a complex state structure. In this context, advancing the role of the market requires different social and political practices. At the moment, it does not mean a rising market society or societal self-organization.

The Chinese experience also prompts us to rethink Harvey's notion of urban entrepreneurialism. Rather than applying his theory, re-reading Harvey gives some new insights. He reminds us that neoliberalism is a set of strategies for 'class restoration' in response to the welfare state, which reduced the profit of the capitalist classes (Harvey 2007). When he participated in a colloquy in 1985, he witnessed the formulation of this strategy, lending support to the private sector.[13] Later, a wide range of operational tactics, including 'place promotion,' was deployed. The state response in this specific setting is 'flexible accumulation.'

Along with this class restoration perspective, economic reform and open-door policy mean a strategy for restoring a capital-owning class cultivated during the period 1949–78. Unlike the challenge of the welfare state, its position is challenged by the internal political turmoil of the Cultural Revolution and comparatively slower paces of economic development than those of the newly industrializing economies in East Asia. This reflects the regime's continuity of economic reform, as they were not to overturn their class interests.

However, actually existing neoliberalism includes a wide range of policy practices, strategies, and tactics (Brenner and Theodore 2002). Harvey's entrepreneurialism thesis is developed from a political economy analysis grounded on post-industrial Western economies. The crisis also allowed the Chinese state to implement its strategies under neoliberalism, which does not imply a waning role of the state. Recent studies argue that urban entrepreneurialism has varieties (Phelps and Miao 2020). One salient feature is municipal statecraft beyond reactionary neoliberalism (Lauermann 2018). China's state entrepreneurialism resembles financialized statecraft—a strong managerial calculation and maneuvers to maintain the government system (Pike et al. 2019).

From this understanding, the entrepreneurial city put forward by Jessop and Sum (2000) is not an application of Harvey's urban entrepreneurialism, as mentioned earlier, because they stress the role of the state. Their entrepreneurial city differs from urban entrepreneurialism in that the entrepreneurial city represents 'inter-scalar strategies.' These studies demonstrate a spectrum of possible strategies, ranging from being more hands-off to leaving the development to the growth machine and more state actors participating in the market responding to external and internal crises. One commonality is their dynamism with the operation of the market. Despite an outlier of governance practices, the Chinese state using the market as instruments might not be so exceptional (Robinson et al. 2022), as seen from the state-theoretical perspective (Jessop 2002b). These studies are not too dissimilar to what was in Shanghai in the 1990s, and perhaps even more so in the 2000s under the world-factory regime (Wu 2003b).

Urban entrepreneurialism describes a phase in which the capital controls the urban process. Now, however, structural conditions have changed. There is a global tendency of rising state capitalism, and the toolkits of urban entrepreneurialism have been greatly expanded. In this context, urban entrepreneurialism is claimed to have ended. In response to this thesis, we can see that state entrepreneurialism utilizes a diverse set of toolkits. Some are entrepreneurial, while others are more directly involved in the economy. State capitalism is not equivalent to neoliberalism; at the same time, however, like neoliberalism, it is subject to capital accumulation. It is an outcome of intertwined capital logic for accumulation and territorial logic to maintain the stability of accumulation. As a sovereign nation-state representing a territorial logic, China reacts toward the rising capitalist world under globalization. It also actively contributes to the capitalist world by participating in the global economies. At the same time, it has to address various crises emerging from capital accumulation. Its urban governance is thus simultaneously subject to the capital logic of accumulation and the territorial logic of the multi-scalar state.

Notes

1 Note that the new public management does not necessarily refer to power devolution to society. The aim is to enhance administration efficiency using 'entrepreneurial' methods to reform public management, including public–private partnerships, competitive tendering, and outsourcing. It is also connected with urban studies literature in Phelps and Miao (2020).

2 These models are discussed in detail in Pierre (2011).

3 The post-political theory is also studied in planning (Allmendinger and Haughton, 2012) and is subject to debate (Beveridge and Koch 2017).

4 See a later discussion on 'class restoration' by Harvey (2007).

5 The term 'flexible accumulation' was used in the original paper, 'From managerialism to entrepreneurialism' (Harvey 1989), as the state shifts from managing the economy to adopting a new 'flexible accumulation' strategy, see Harvey (2016, p. 158).

6 Harvey (2016, p. 157) reflects, in a commentary after the reprint of his 1989 article on urban entrepreneurialism, 'Michel Foucault, … [h]is ideas played no role in constructing the transition I describe but they do provide an interesting interpretation of the potential and highly negative political consequences of changing political subjectivities. … In 2005, in *A Brief History of Neoliberalism*, I rewrote the story of the political-economic shifts behind the rise of urban entrepreneurialism under neoliberalism.' We argue that this re-writing brings his interpretation slightly away from governance, triggered by 'flexible accumulation,' and closer to governmentality interpretation, i.e., a deliberate art of state governing.
7 Thanks to Yi Feng for suggesting this point.
8 See also Miao and Phelps (2019) for the study on Singapore.
9 Here, the 'political' means the struggle over social equity. For a brief introduction to the concept of the post-political city, see Davidson (2019); for the original conceptualization, see Rancière (1999).
10 In a similar vein, but with a different conclusion, Wu (2022a) describes how a visible state emerges from the construction of various neighborhoods in China.
11 Mainland China imported this land model and created its land-finance regime.
12 The remaining of this paragraph is adapted from Wu (2018b, p. 1384).
13 This is mentioned in Harvey (2016, p. 134).

2 State entrepreneurialism

Historical formation and practices

The next section follows Chapter 1 on governance as a mode of regulation and explains its emergence from China's political economy. Rather than thinking of governance as a given ideology imported or imposed by the state, the chapter explains how it has become a specific statecraft to cope with various challenges in historical moments.

The historical formation of China's entrepreneurial statecraft[1]

This section provides an account of the history of China's statecraft and explains how 'governance' emerges as a toolkit, which generates the specific form of 'state entrepreneurialism.' Hence, our understanding of state entrepreneurialism is historical, grounded, and generative instead of arising from theoretical deduction. The description of historical moments serves to illustrate how governance changes are triggered by specific state territorial actions in crisis management and concerning changing global political economies. The historical review aims to demonstrate that state entrepreneurialism is not a theoretical construct but rather a phenomenon of historical materialism. It is, in this sense, a 'grounded' product.

Constrained urbanization under the centrally planned economy (1949–1978)

Entrepreneurialism emerges within the contradictions of state socialism. From 1949 to 1978, during the Cold War, China adopted an import-substitution economic strategy and developed its economy through self-reliance. Investment in heavy industries became a priority. The state owned the means of production, and the labor market was absent. The state had to organize collective consumption in cities while separating the urban and rural sectors. The surplus was extracted from the rural sector and devoted to supporting the state industrial sector. China maintained a very low level of urbanization. The percentage of the urban population was kept below 20 percent of the total population until 1978 (Zhou and Ma 2000). China was under-urbanized with a clear urban–rural divide compared with its industrial capacity. This under-urbanization shares some similarities to socialist economies in contrast to the Third World. This is a unique feature compared with the Western industrial world, which started with land enclosure, industrialization, and urbanization.

The strategy of 'economizing urbanization' (Chan 1994) and its corresponding urban process—not investing in urban development—led to its own contradiction of capital accumulation (Wu 1997). Urban consumption, though necessary, was 'unproductive' in this model. Investment in urban development could not generate a surplus or, in market terms, overcome low profitability. There was a lack of effective demand, though the need for social reproduction was acute in the 'shortage economy' (Kornai 1980). Under the import-substitution

DOI: 10.4324/9781003355175-3

strategy, the state suppressed the cost of labor and, consequently, consumption. Capital accumulation relied on production, whereas consumption did not drive economic growth. The rural sector saw a large idle labor force, and the urban state-owned industry was inefficient. The idle labor could not be put into production because the policy of import-substitution industrialization pursued at this time was capital-intensive and did not absorb the labor force in large quantities. Because of the household registration system, labor was not commodified and immobile. This shows a stark difference from Western urbanization processes.

The statecraft was administratively centered or 'centrally' planned to manage this form of political economy. Despite the significant difference from the welfare state, the lack of further scope for expansion or 'accumulation space' in the production sphere is remarkably similar to the over-accumulation crisis under capitalism (Wu 1997). This explanation relates Chinese urbanization with the more general urban process under capitalism (Harvey 1978). It also indicates a possibly similar trajectory toward entrepreneurialism after the crisis of the welfare state in the West and the planned economy in China.

Initial market-oriented reform and the embryonic market economy (1979–1991)

In the 1980s, China saw liberal components of market regulation inserted into its rigid, centrally planned economic system (Huang 2008). This was not intended to replace the state planning system with the market as a governance principle. Rather, the market was brought in alongside planning to form a dual-track system. In other words, the market was not seen as a governance principle for the whole economy but rather as a new space to expand accumulation. This market turn has brought a supplementary space of accumulation. This has triggered 'local development corporatism' (Oi 1992).

Starting with rural reform, cadres engaged in the business of township and village enterprises (TVEs). Market-oriented reform was more successful in rural areas. Rural de-collectivization through the introduction of the 'household responsibility system' significantly increased the incentives and productivity of agricultural production. In short, entrepreneurial activities and entrepreneurialism emerged outside the formal state space in rural areas. As the market price was higher than the price of state-allocated resources in such a dual-price track system, some officials used their power to sell their controlled resources in the market for a profit, leading to corruption. The regulatory context also tended to lead to rent-seeking by the predatory local state. To solve this problem, more aggressive price reform was initiated in the late 1980s, but it triggered hyperinflation and eventually political and social contention in 1989.

Throughout the 1980s, the development of the urban sector expanded the space of accumulation but did not evolve on such a scale that it could overcome its constraints. The fundamental contradiction was low purchasing power. While entrepreneurial activities prevailed with industrialization, the governance principle did not shift toward urban entrepreneurialism. According to Jessop and Sum's (2000) definition of the 'entrepreneurial city' in a Western context, the market development in this phase created entrepreneurial activities in the city but not the entrepreneurial city—a changing territorial governance form. The city is a container for entrepreneurial activities, but the urban politics are more managerial.

The historical conjuncture moment is symbolized by the relocation of labor-intensive manufacturing industries from developed economies in the West and newly industrializing economies in East Asia to China. Economic development was fostered mainly by regulation exceptions given to rural families and villages to initiate new enterprises, and more exceptions were extended to local governments to build entrepreneurial cities after Deng's Southern China tour in 1992.

Export-oriented industrialization and emergent urban entrepreneurialism (1992–2000)

Deng Xiaoping's Southern China tour in 1992 marked a significant shift in capital accumulation. With the inflow of foreign capital, the local state had to offer preferential treatment to external investors to attract investment, similar to US local governments. The influx of foreign investment and inter-city competition made governance more market-friendly. In the 1990s, China initiated a series of market-oriented reforms in labor, land, housing, and social services, which could be regarded as the road to capitalism (Walker and Buck 2007). This means a more prominent role for a market economy, while capitalist ideology is constrained.

The marketized rural economy suffered as many TVEs went bankrupt, pushing rural laborers into the urban market. Under competition from foreign investment, state-owned enterprises (SOEs) underwent large-scale bankruptcy after the mid-1990s, and cities experienced consequential welfare retrenchment as the 'iron rice bowl' or welfare net associated with SOEs was dismantled (Solinger 2002, 2018). Development zones became widespread, where central planning mechanisms were suspended by special policies and market exchange prevailed. With rising market opportunities, government officials directly participated in profitable sectors such as real estate. They developed coalitions with external investors in production activities, initially referred to by Duckett (2001) as 'state entrepreneurialism.' This earlier version of state entrepreneurialism focuses on the state's economic involvement as a profit-making activity. It does not refer to the feature of governmentality—the art of governance—as the central core of state rationality.

This differs from using regulatory power to extract rents in the earlier dual-price track system, as marketization prevailed in the Chinese economy. However, this notion of state entrepreneurialism does not mean state strategic policy guidance, as suggested later by Wu (2018b). It stresses the state's participation in market activities without distinguishing its motivation from the private sector. It regards the intention as profit-making or generating local revenue as well as personal benefits and corruption. In the literature, urban entrepreneurialism is placed within the context of circulating neoliberal policies (Brenner and Theodore 2002; McCann and Ward 2011). State entrepreneurialism does not regard its entrepreneurialism as originating from policy mobility. Instead, it has a similar profit motivation. Rather, the means of profit-making changed, involving a shift from earlier rent-seeking behavior, for example, imposing a fee on TVEs because the local state was able to restrict their activities, to a more productive market engagement, for example, providing cheaper land to external investors and forming a coalition with the private sector in economic development zones.

China's market-oriented reform in this period was once again closely related to world-historical moments. The initial opening and development of the Pearl River Delta were closely associated with the newly industrializing economies in East Asia, especially Hong Kong. These nationwide reforms were driven by major global conjunctural moments. For example, the Asian Financial Crisis in 1997 triggered China's housing commodification in order to promote domestic consumption. The system of public housing provision was abolished. The actual measures of housing privatization were, in fact, more dramatic than those recommended at the time by the World Bank. Since then, new housing has been predominantly developed through the market, which triggered widespread urban demolition in the late 1990s, continuing into the early 2000s. In this period, China applied the toolkits of neoliberalism, such as welfare retrenchment, housing commodification, and the establishment of development zones, and also experienced its symptoms, such as excessive encroachment of rural land, informal settlements (known as 'urban villages') and urban poverty experienced by laid-off workers. The neoliberal reform introduced by the nation-state in the 1990s prepared China to join the World Trade Organization (WTO), introducing a new global conjunctural moment.

The world-factory model and the 'double movement' (2001–2007)

China's joining the WTO in 2001 marked the maturation of export-oriented development. China had become the world's factory, linked to the global conjuncture through changes in the new international division of labor (NIDL). The rapid expansion of export markets led to the influx of rural migrants into Chinese cities. Market reforms in the 1990s had led to a competitive pricing of production factors, including labor and land with relatively well-developed infrastructure. Different from the development zone model in the 1990s, the world-factory model was no longer a model of 'state exception' (Ong 2006). The operation of the world factory requires more wide-ranging state assistance and intervention, as shown by 'planning for growth' (Wu 2015b), in which the local state organizes land acquisition and leveling and infrastructure provision through its development corporations (Hsing 2010; Lin 2014; Liu 2019; Feng et al. 2022a). In this approach, state involvement in development is remarkably different from neoliberal planning, as the state is more directly involved.

The state plays a critical role in maintaining the structural coherence of accumulation. Facing increasing social tension and the crisis of social reproduction, the leadership of Hu Jintao and Wen Jiabao, starting in 2002, shifted the policy from welfare retrenchment through laying off state industrial workers in the 1990s to establishing a basic social security system. Its development approach changed from Deng Xiaoping's 'growth-first' to 'scientific development.' This is not merely rhetoric but reflects a version of a Polanyian 'double movement'— a dialectic process as the movement to expand the scope of the market into the society, a protective countermovement emerges (Zhang 2013), involving a suite of policies to respond to social tensions and support labor through labor protection, the development of a socialist countryside, basic agricultural land and environmental protection.

The state strengthens its regulatory power as a result of the double movement but also due to the problem of rampant land development. The earlier version of urban entrepreneurialism of the 1990s did not last long due to the contradictions, social intentions, and urban environmental problems it created (Xue and Wu 2015; Wu 2016a). The state strived to develop a competitive land market through open auctions and bidding to promote land market transparency. This, in turn, fostered a more market-oriented land development (Xu et al. 2009; Zhu 2019; Qian 2022). However, at the same time, the state could impose greater control over the quantities of land development, for example, through the linked land policy (Ong 2014; Tian and Guo 2019; Shi and Tang 2020). In this period, Chinese cities, as the global factory base, managed to build massive infrastructures and extensively integrated into global production networks (GPNs). With the booming real estate market, Chinese cities also experienced rapid suburban development and the redevelopment of traditional neighborhoods (Wu 2022a).[2]

Financialized state entrepreneurialism (2008-present)

In the world-factory model, the state's role has been strengthened. However, the changing global political economy further required the state to enhance its capacity for crisis management through financializing housing (He et al. 2020; Wu 2022c; Wu 2023a). In response to the global financial crisis in 2008, the Chinese government initiated a fiscal stimulus package of 4 trillion Yuan. The stimulus package did not allocate fiscal revenue to local governments, but it did allow them to secure debt financing (Wu 2023a; Bai et al. 2016; Naughton 2020). Still, the stimulus policy was not strictly fiscal because local governments did not act directly through taxation and fiscal expansion. It is indirectly achieved through financial borrowing of state-owned enterprises (SOEs) and their involvement in urban development under local government control. This was more like a financial program and different from the Keynesian approach of taxation, welfare provision, and collective consumption.

The stimulus package included investment in railways, highways, airports, affordable housing, rural infrastructure, technological innovation, and environmental projects. In short, the package relied on infrastructure-led development. Local governments had to offer matching capital to get credit through the stimulus package to bring forward infrastructural investment. To find ways to raise capital for co-investment from the capital market, the local state borrowed from banks through various conduits, often off the banks' balance sheets, for example, through 'wealth management products.' The stimulus package casts a long shadow of financialization (Wu 2023a).

State-owned banks dominate the capital market. Financialization aimed to leverage savings deposited in banks into investment, as China has a high savings rate (Wu et al. 2020). Local investment platforms were established to mobilize capital, forming China's debt-driven urbanization (Tsui 2011; Pan et al. 2017; Feng et al. 2022b). The earlier development corporations were readily transformed into local government financial vehicles (LGFVs) to raise capital (Feng et al. 2022b; Wu 2022c). These corporations are SOEs owned by any entity within the government hierarchy. The state is central to this financial operation (Wu 2023b).

The financialization approach means that local states no longer prioritize competing to gain external investment but instead use their own financial vehicles to raise capital.[3] As a result, urban development politics in this phase became more state-centered, characterized by 'financialized state entrepreneurialism' because the local state uses financial approaches to mobilize investment capital for development (Wu 2023b), for example, the government-investment fund (Pan et al. 2021). It does not rely on capital inflow from the global financial market. Land development after the global financial crisis is not about profitability, but rather about state leverage of land ownership and the assets of state-owned enterprises for development finance (Wu 2022b). This financial approach strengthens rather than weakens the state's role (Wu 2023a), in contrast to financially dominated mega-urban projects in Europe (Swyngedouw et al. 2002) and austerity urbanism (Peck and Whiteside 2016, Peck 2017b).

Defining features of state entrepreneurialism

The notion of 'state entrepreneurialism,' meaning a more strategic state role in urban governance, originated from grounded observation of Chinese urban development.[4] In contrast to American 'edge cities,' large-scale peripheral developments in China represent first the state participating in the development through various development agencies and arms and second strategic guidance toward achieving a territorial logic beyond the locality. The prototype is China's development zone. Here, we have a very different interpretation from the 'special economic zones,' where its governance is seen as neoliberal governmentality—exempting some sovereignty of the nation-state.[5]

For example, Yizhuang, a new town in Beijing, is a planning outcome of state entrepreneurialism (Wu and Phelps 2011). In the original place of the rural town of Yizhuang, the Beijing Economic and Technological Development Zone was established (Figure 2.1). The figure shows a different set of spatial boundaries of administrative hierarchy (in this case, the rural town of Yizhuang under the Daxing district) and a development zone superimposing on this jurisdiction system. This contrasts a spontaneous clustering of office and high-tech firms attracted by a business-friendly local environment or entrepreneurial urbanism in the post-Fordist flexible accumulation; the new town has been developed through state planning, although the planning

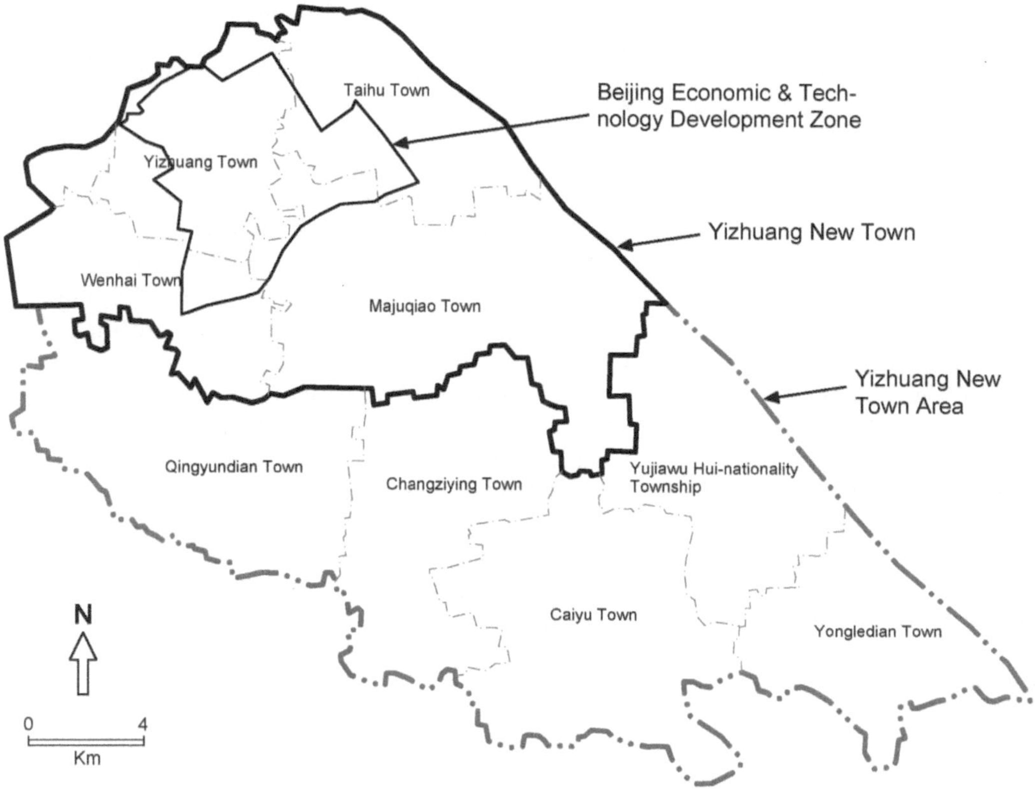

Figure 2.1 Yizhuang New Town and the Beijing Economic and Technological Development Zone.

Source: Wu and Phelps 2011

process solicited foreign consultants (Wu 2015b). The new town is an upgraded version of industrial parks and development zones with mixed uses. Within the boundary of the designated zone, the rural areas were re-assigned to the control of the state development agency, which controls the land and forcefully acquires the land for urban development. At the same time, existing town governments take responsibility for social management. Yizhuang has been developed by the state development agency as Beijing municipal government's development arm.

From this example, we come to the definition of state entrepreneurialism. This refers to the 'state apparatus, in particular the local state, demonstrating a greater interest in introducing, developing and deploying market instruments and engaging in market-like entrepreneurial activities' (Wu 2018b, p. 1384). Here, the state directly participates in market activities through various mechanisms, such as state-owned enterprises acting as development corporations. These state-owned enterprises are very complex.

Nevertheless, entrepreneurialism is only one component. It is more or less used as an instrument to achieve state intention and maintain its centrality, as the definition further clarifies that a 'series of state entrepreneurial actions to fulfill its strategic intention to maintain economic growth, stability, and capital accumulation and, in turn, its governance capacity' (Wu 2023b, p. 365). The instrument of state-owned enterprises is driven by complex motivations, not just for profit maximization. Nevertheless, these enterprises must follow the market mechanism as a market agent. The state is also multi-scalar and has varied motivations and objectives at different levels of operation. The central government may be more concerned about financial risk and social stability, while local governments are more interested, for example, in revenue generation in land development.

From the observations of two components: the state and entrepreneurialism, we can summarize two defining features of state entrepreneurialism.[6] First, the state acts *through* the market as an initial version; later, however, it also acts *in* the market in the second version of state entrepreneurialism. The notion of *through* the market may imply a governmentality approach—'governing at a distance' following the market logic. In the literature, the notion is also called the state in market, as the state only operates as a regulatory state.

Acting *in* the market differs from neoliberal governmentality. The state not only creates a market environment or a market mentality for governance (hence, governmentality) but also directly controls or participates in the public sector in the market. In the literature, the notion is also 'market in state,' meaning that the state absorbs market coordination into its own operation (Zheng and Huang 2018). Here, we stress the intertwined nature of state and market. State entrepreneurialism means the state actively engages with the market through quasi-government agencies.[7] In China, these agencies include urban development and investment corporations, land reserve centers, various state-owned enterprises, and financial instruments such as urban development corporation bonds and local government bonds. This is the second version of state entrepreneurialism as the state acts *in* the market.

Second, the state also reflects a territorial logic. The state acts directly in the market and is positioned at the center of making strategies, characterized by 'planning centrality.' While market instruments are utilized, state strategy is placed at the center of governance. Hence, there is a contrast between 'planning centrality and market instruments.' Planning centrality reflects a more strategic role of the state beyond 'reactionary politics' as a critique of municipal statecraft toward urban entrepreneurialism (Lauermann 2018).[8] It has a strong intention beyond market rationality, and in the case of China, it is the communist party's development goal. To achieve the development and governance intention, the approach to governance must be flexible and market-friendly. The state also uses market actors or introduces a market mechanism into development. This is state entrepreneurialism version 1.0, namely introducing an existing market actor. Planning centrality indicates that the state is not overwhelmed or captured by the market or financial logic. The state does not give up its overall control. However, it has to respect the market logic in order to implement development projects and govern society.

For example, utilizing a financial approach, the state creates a private equality fund, such as a government-guided investment fund, to target a particular industry (Pan et al. 2021, also see Chapter 4). Market reform is achieved through technical terms rather than ideological forms such as neoliberalism. The state welcomes investors not just in the priority sector or area but also through land value capture and collateral to raise development finance through its agents. This technical change is different from ideological change. That is, the rationality and principle of market exchange are not strictly followed for its strategic intention. Nevertheless, it is this 'violation' that lays down the contradiction, which, in a different historical moment this contradiction may lead to another crisis.

Governing urban development under state entrepreneurialism

State entrepreneurialism provides an overall governance framework for planning and development. In the following, we discuss three policy priorities in China: urban regeneration, suburban development, and rural revitalization.

First, *urban regeneration* has become a policy priority as Chinese cities face tighter land control after rapid urban expansion. We will discuss detailed policy practices in Chapter 4. Here, we explain the trajectory of urban regeneration, especially the so-called 'three old' (old villages, old urban areas, and old factories) regeneration and micro-regeneration, to illustrate how they embody state entrepreneurialism. In the 1990s, property-led redevelopment accompanied by large-scale demolition demonstrated a feature of 'neoliberal' urbanism. However, the local state facilitated residential relocation at the time instead of just reactionary toward real estate developers (He and Wu 2009). The land market has been formalized through the compulsory requirement for public auction since 2004. Urban demolition became ubiquitous and led to widespread social discontent. The global financial crisis also severely constrained land development.

In Guangdong, the former party leader negotiated a special policy from the central government to allow the 'stakeholders'—existing landowners—to lead the redevelopment. This institutional innovation used social and market actors to speed up urban regeneration. The redevelopment did not require the sale of land to the municipal government to become a releasable land into the 'open' market. Instead, existing landowners could find developers and financial sources themselves. This 'informality' led to the boom of market-oriented and self-organized redevelopment. It significantly stimulated the development of urban villages.

However, the development is not entirely spontaneous and bottom-up. A special urban redevelopment office must approve the redevelopment projects under government programs. Indeed, the government has forgone land profits and relaxed planning control. The actual development divides land into three parts: *in situ* rehousing of villagers; a real-estate project to generate development finance; and a plot of land reassigned to original villagers as the collective assets (Wong et al. 2024; Liu et al. 2023). These regenerations reflect the state's intention to involve market actors to regenerate urban areas and, in turn, create a new space for economic development.

Later, facing increasing social contests over urban demolition and the damage to heritage, urban regeneration stopped. The new policy required villagers to reach a consensus before demolition. Hence, it requires greater 'social participation.' Since the mid-2010s, a new policy experiment called 'micro-regeneration' has been recommended (Wang et al. 2022a, 2022b). The micro-regeneration paid more attention to job creation through mixed land development. The intention of this kind of regeneration ranges from coping with the problems of unfinished demolition and redevelopment programs to responding to new political mandates of the central government for heritage preservation and 'making people happier' (Wu et al. 2022b). The agenda has recently shifted to 'social governance,' especially 'co-production' (*gongtong tizhao*) under 'Cities for People' (Li 2022).

The changing course of urban regeneration reveals strategic guidance of a multi-scalar state. Urban regeneration has become more strategic, fulfilling the central government's political mandates or local government's metropolitan-wide initiatives. This is particularly true when urban villages are located at the strategic development site. For example, Guangzhou wishes to develop a second business center—the Guangzhou 'financial town' after the successful development of its CBD in the Pearl River 'new town' (Figure 2.2) (Wu 2015b). These urban villages entirely vanished. Similar to other mega-urban projects discussed later in this chapter as examples of state entrepreneurialism, China's urban regeneration reflects a new mode of governance that uses the market instrument but at the same time maintains planning centrality. The project aims to restructure Guangzhou's economic structure to strengthen the financial sector. Regeneration is a carefully chosen strategy to achieve a more strategic goal;

Figure 2.2 The central business district of Guangzhou and its financial town.
Source: Photo by Yuqi Liu.

existing villagers are rehoused, and some are compensated. This is not market gentrification; instead, it brings the state to develop a new economic space, such as a financial center.

Second, *suburban development* represents the state's strategic guidance over peripheral spaces to upgrade the economy and make a more sustainable urban future. It represents the state's effort to address earlier urban sprawl driven by real estate development. Because of the greenfield location, development projects are large-scale. Suburban development is organized through new towns and mega-urban projects. Through policies such as transit-oriented development, suburban development concentrates on the area around the mass transit and creates large settlements. The development is usually not initiated by local communities. The overall development objective is justified by the central government initiatives, such as an 'indigenous innovation nation' and 'ecological civilization,' especially for major development in the large cities, as we see the development of greenways in Chengdu (Zhang et al. 2022) and Zhangjiang Science City in Shanghai (Zhang and Wu 2012; Zhu et al. 2023). The municipal government plays a key role in formulating urban master plans and visioning the spatial structure of city regions. Still, the implementation is achieved through a combination of the management committee as the government authority and development corporations. The governance of mega-urban projects reflects state entrepreneurialism, which will be discussed in detail later in this chapter.

Third, *rural revitalization* represents the state's efforts to address 'three rural questions'—the peasant, agricultural, and rural questions. China's spatial governance presents prolonged urban–rural dualism. The policy started with the development of the socialist countryside in the 2000s and was upgraded to rural–urban integration, which was piloted in Chengdu (Ye and LeGates, 2013). This means rural areas are no longer treated as separate entities but are linked to the same market-oriented urban process. Rural development has become widespread, presenting a feature of arguably 'planetary urbanization'—urbanization occurs at the global or planetary scale (Brenner and Schmid, 2011).

This has significantly changed the rural landscape. Public finance in rural areas has been weak (Po 2011). In the Pearl River Delta, farmers' income relies on land rent and rural collectivized areas. However, the global financial crisis in 2008 led to the decline of export-oriented industries. The crisis of local public finance was particularly severe in the city of Dongguan. Since then, the government tried to consolidate scattered rural industries into new industrial parks. For example, it has successfully attracted the relocation of Huawei R&D facilities from Shenzhen. For example, the state has strengthened its management of rural areas through land quotas. Farmers' houses in remote areas were demolished to generate new quotas for development near the city. The local government also pays greater attention to basic public facilities in rural areas. For example, Jiangsu provincial government initiated the village improvement program. This mainly aims to improve the physical environment and public hygiene, watercourse maintenance, waste collection, animal manure recycling, safe drinking water, landscaping, and preserving cultural heritage.

The policies are implemented through projects using the market instrument, as the government has limited resources. However, these projects are often initiated by town or street governments. Rural revitalization involves multiple actors and diverse motivations. For the local state, the projects demonstrate a beautiful rural landscape, following the mandate of 'Beautiful China' and various political discourses (Wu et al. 2022b), which might not be profit maximization. On the other hand, as shown in the village improvement program, it is not an entire welfare project providing free housing to low-income farmers. Rather, the multi-scalar state demonstrates its multiple and sometimes contradictory objectives.[9] However, market actors also consider the market logic, and their participation may, in turn, overthrow the original intention. As a result of state entrepreneurialism, rural areas have become formalized. Rural areas present features similar to urban areas, such as large-scale development zones. It is the urbanization of the countryside and aestheticization of rurality.[10] Traditional rural collectivism has been dismantled, varnished, or utilized, but the governance principle is a mutated private form (Wu 2022a).

Mega-urban projects—examples of state entrepreneurialism

In this section, we use mega-urban projects to demonstrate the application of state entrepreneurialism. This dialogue is mostly with urban entrepreneurialism literature to reflect the state's strategic and territorial intention. On the other hand, delivering mega-urban projects involves institutional reconfiguration for great market efficiency, as described by new public management or 'intrapreneurialism' (Phelps and Miao 2020). We argue that the key difference is their treatment of market rationality. Although this can be seen as introducing the market into the state (specifically, financial logic when using financial products), the state is above market principles rather than using the market to restructure the state. The state, as a social relation, is situated in capital accumulation. Its rationality is not transformed by a new public management theory or, even more widely speaking, policy mobility—transferring policies across jurisdictions.

Large-scale urban development projects, especially waterfront regeneration, are often interpreted as an example of neoliberal urban governance (Swyngedouw et al. 2002). The regeneration of London's Dockland and Canary Wharf as a financial center arguably shows a limited role of the state, attracting overseas property developers by reducing its regulatory intervention. In reality, however, providing infrastructure such as the DLR and Jubilee Line was critical for the eventual survival of Canary Wharf as a business center after the developer's bankruptcy. Many mega-urban projects are located in the peripheral areas. Large suburban developments such as edge cities represent the local growth impetus under neoliberal urbanism (Peck 2011). However, from mega-urban projects in China, we can detect a different mode of governance—state entrepreneurialism.

Although China has seen formal and informal developments in the peripheral urban areas, mega-urban projects emerged as a new formal development approach. In the earlier stage of market reform, informal development was a major approach. For example, in the city of Shantou in Southern China, small workshops of township and village industries mixed with rural villages and residences for migrant workers created a scattered development pattern. This pattern is widely seen in the Pearl River Delta during export-driven industrialization (Sit and Yang 1997). This form of semi-urbanization resembles the 'desokata' model with mixed urban and rural uses (McGee and Robinson 1995).

However, in suburban Beijing, the Beijing Economic and Technological Development Zone, a large-scale planned development, has evolved into an entirely new town at Yizhuang. The Yizhuang New Town has been developed under the master plan. While industrial zones in rural China use the land inefficiently, the new town adopts a compact form of development. Besides industrial land, it also accommodates commercial and residential uses. Thus, it is beyond an industrial zone. The land is developed according to zoning and the land use plan. The new town itself is a mega-urban project.

Other examples include Zhengdong New District and Wuxi Taihu New Town. Zhengdong is a suburban town built into Zhengzhou's new central business district. Wuxi Taihu New Town is the new site for the Wuxi municipal government and modern industries. It is also built through the ecological fix of polluted land in the era of TVEs (Zhang and Wu 2022b). They are planned large-scale urban development, often in the former rural area (Figure 2.3). The picture shows the water landscape in southern Jiangsu. The sharp contrast between the rural and urban landscapes indicates the dominance of the urban—a process of planetary urbanization, different from the desokata model of spontaneous rural industrialization in the Pearl River Delta.

China has seen the development of diverse new towns (Shen and Wu 2020; Wang 2022). 'University towns' are a special type developed based on university campuses. They have become an attractive form of large-scale urban development. Universities are drivers for these developments (Shen 2020). In the 2000s, universities significantly expanded student recruitment, requiring more space. The original campuses in the central area are too small.

Large Chinese cities have expanded very rapidly. They are faced with tighter land management from the central government, and the municipal government wished to develop a more compact form. In the early stage of land development, education use in suburbs was exempted from land auctions, and, hence, the universities could obtain cheaper land from the municipal government. The district government used this opportunity to collaborate with universities to build university towns. The universities began to use land to finance their new campus development.

Previously, suburbs were underdeveloped and scattered with real estate projects in the form of urban sprawl. To increase the attractiveness of the suburb, the local government leased suburban land to universities at a lower price. This development did not directly burden the local government because the university town was financed by universities who borrowed development funds based on their projected student numbers. Through the development of

Figure 2.3 Rural landscape of Wuxi in Southern Jiangsu.

university campuses, the suburban new town attracted more residents, which helped fund infrastructure and services development. Therefore, developing university towns is an effective way to stimulate suburbanization. The town is designed through a master plan for an instant new image.

The main problem of university towns is that they are not walkable compared to central urban areas' campuses. The grand design does not pay sufficient attention to walkability, perhaps following the model of large-scale new development. In reality, the development often lacks coordination. University towns are built more like development zones than urban areas, lacking social interaction and common facilities beyond individual universities. However, some began to see more urban life (Liu and Yau 2020). These large-scale developments show that suburban development in China has gone beyond suburban sprawl (Miao and Phelps 2022). Compact developments such as university towns and other transit-oriented developments (TODs) are encouraged, all well-planned and designed, reflecting the state development intention.

The mega-project of Lingang New Area in Shanghai

An example of large-scale urban development is the new town of Lingang in Shanghai.[11] According to the master plan, it is one of the nine new towns in Shanghai, and it uses a polycentric urban spatial strategy—'one city, nine towns'—as proposed in 2001. The new town is located about 75 kilometers from the city of Shanghai. It is an entirely new city that combines industrial and residential uses. In 2001, following the construction of the Yangshan deep-water container port, Shanghai perceived an opportunity to develop advanced manufacturing industries, which initiated the development of Lingang. A heavy equipment manufacturing zone was set up. A new town near the manufacturing zone was planned to support industrial development. In 2002, a German-based architectural firm, German, Mark, and Partners (GMP), won the competition for the conceptual master plan and designed circular zones surrounding a large lake. The 'garden city' concept inspired the design of circular zones.

The development of Lingang was initiated after Yangshan developed into Shanghai's deep-water port. The port allows the modern container ships to pass through. The port is the largest in China, with a high standard of receiving the most advanced container ships. A significant proportion of Lingang was reclaimed from the sea. Lingang is connected to the city of Shanghai by a metro line. Three new stations with a modern design are inside the Lingang area. These stations are under-used because of the distance from Lingang to central Shanghai. The distance is beyond daily commuting, but many working in Lingang still live in central areas. Outside the metro station is a large sunk square. The design standard is high, but there are few users for most of the day. The station at Shuyuan, a modern metro station installed with an elevator, is in the middle of rural areas, not far from the farmers' houses and agricultural fields. The Lingang new area shows a mix of formal mega-urban projects and informal rural settlements (Figure 2.4).

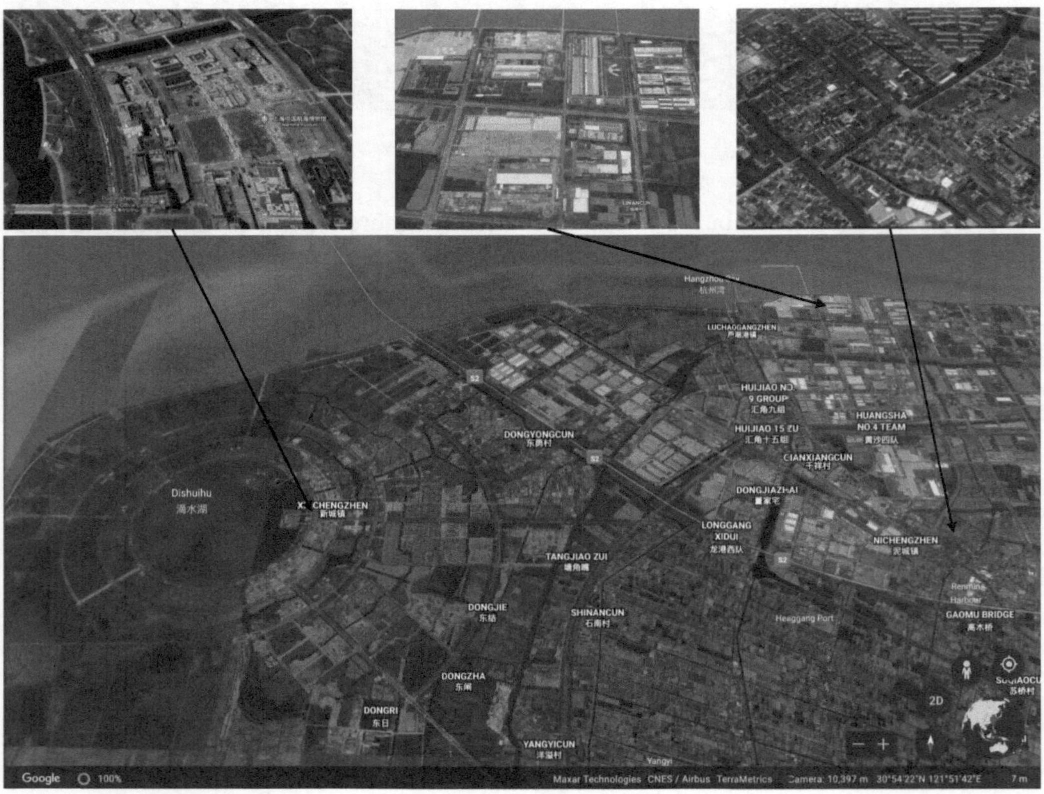

Figure 2.4 Lingang New Area with planned industrial zones, the new town center, and rural towns and villages in Shanghai.

This development has a strategic purpose. When the deep-water port in Yangshan was developed, the Shanghai municipal government regarded it as a good opportunity to upgrade its manufacturing industries from the traditional labor-intensive sector, which has been relocated to nearby rural areas outside Shanghai. The new town's development hoped to support the deep-water port (Figure 2.5). A new town was thus planned to benefit from the synergy between the port and the city. It was believed that, at that time, heavy equipment manufacturing industries might be relocated from developed economies in the West. The development of the large industrial zone would help Shanghai secure its manufacturing status after severe de-industrialization and the layoff of workers in state-owned enterprises.

A heavy equipment manufacturing zone was built at the start of Lingang. The mega-urban project contains several functional areas, a logistic zone, and a new town center for residential and commercial uses. The whole area is very large, comprising an area of 315 square kilometers, and has now expanded into 873 square kilometers, equivalent to the size of Singapore after the establishment of Lingang New District (*pianqu*). Because the land of the new town center was claimed from the sea, the government controls land ownership. In Lingang, Shanghai has the last large parcels of land for large-scale industrial enterprises. In the jurisdiction of the Lingang new area, there were four new rural towns. One rural town, Lucaogang, was absorbed into the urban jurisdiction. Because the distance between the industrial zone and the new town center is too far, involving a drive of half an hour, another rural town,

Figure 2.5 The deep-water container ship port of Yanghshan, Shanghai.

Nicheng, has seized this opportunity and developed shopping malls. It provides convenient services to manufacturing areas and is transformed into a small town with a prosperous urban life. The development eventually transformed the landscape of the former rural town.

The modern industrial manufacturing zone was developed effectively, although it did not attract many overseas investors. It did, however, attract the relocation of domestic firms (Figure 2.6). For example, the SANY Group, which specialized in heavy equipment manufacturing, relocated into the zone. Despite slow progress, Lingang also attracted Tesla's Gigafactory to its industrial zone, which covers an area of 84 hectares, making it the biggest foreign manufacturing industries in the city. The project produced 710,000 cars in 2022, more than half of Tesla's total global output, with an investment of $5 billion. Although the development of Tesla was not planned, it is only possible after nearly twenty years of construction of the Lingang New Area.

A major challenge is to attract the population to Lingang to create a vibrant new city. Although industrial development started effectively, the new town was slow to attract residents. Its population in 2000 was 156,000, growing to 212,000 in 2010. The annual growth rate was only slightly over 3 percent. Of these, the migrant population accounted for 79 percent. The target was to achieve a population of 800,000 in 2035. Population growth lagged compared with other new towns in Shanghai, such as Songjiang. The streets in Lingang are still

Figure 2.6 The Modern Heavy Equipment Manufacturing and Logistic Zone of Lingang New Area in Shanghai.

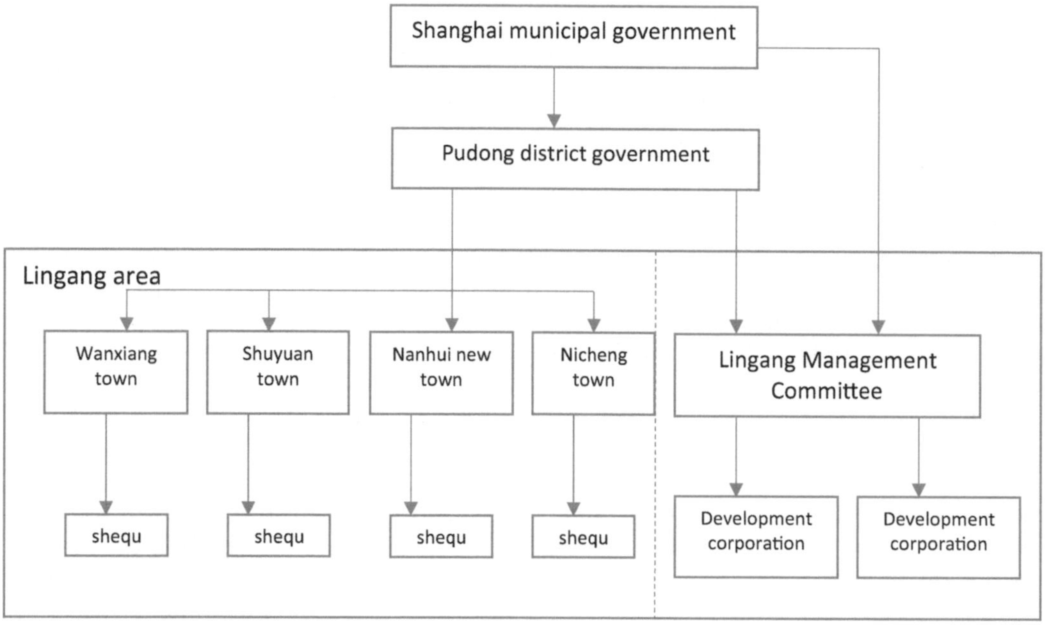

Figure 2.7 Lingang's multi-level governance and development corporations are state agencies operating in the market.

Source: Wu 2022b.

empty, which is common in other Chinese new towns. They are built and planned under a high standard but lack social encountering and urbanism. It is more or less a planned urban development.

In Lingang, development was achieved by eight major state-owned development corporations. They deliver projects ranging from infrastructure to property development. Two corporations are significant. The largest is the Shanghai Lingang Economic Group, which is responsible for the industrial development zone. It now manages an overall development in the whole Lingang area. Another is the Harbor City Development Corporation, responsible for new town development. They are affiliated with different levels of government. The former belongs to the Shanghai municipal government, while the district government owns the latter. In addition to these two major development corporations, four development corporations in Pudong established their local subsidiaries in Lingang at the request of the Shanghai municipal government to support the construction of Lingang, and various other development corporations were set up by joint operation between local and other districts of the city (Shen et al. 2020).[12]

The new town is a mega-urban project, a complex project involving a complicated coronation of multiple actors (Figure 2.7). Lingang's governance is entrepreneurial and managerial, with different levels of government. Although Lingang is a project for Shanghai, the municipal government assigned the management committee to the district government of Pudong. Therefore, the authority of Lingang is 'municipally owned but district-managed' (Wu 2018b, p., 1,390).

State entrepreneurialism as a model of governance

What is the model of governance represented by the large-scale peripheral development like Lingang? Peripheral developments in the metropolitan regions are interpreted as 'edge cities' in the US through the economic dynamics of metropolitan dispersal and local clustering of the post-Fordist economy (Garreau 1991; see also Phelps and Wu 2011). The edge cities are 'spontaneous' clusters out of firm linkages for official buildings, hotels, and new industrial spaces. They are not created by the government but rather out of economic agglomeration at suburban transport nodes. The town government plays only a minimal role in this. The mode of governance is known as 'neoliberal suburbanism' (Peck, 2011), which is driven by entrepreneurial business communities. In the post-industrial era, official buildings clustered in the suburbs, not just in the CBD. Thus, the edge city is a sub-center or business district in the suburb.

Further, a global view of suburban governance shows multiple governance modalities (Hamel and Keil 2015; Keil and Wu 2022). They include the state, capital accumulation, and authoritarian private governance (Ekers et al. 2012). It is not just the dynamics of market agglomeration that are subject to state intervention and land regulation (Storper and Scott 2016). The role of the state is at the core of this 'urban process'—used by Harvey (1989) to describe the process of capital investment in the built environment. The 'state not only as a mediator, but also as a central institution. In that respect, governance does not proclaim the end of the state' (Ekers et al. 2012, p. 413). There are different combinations of modes or a variegated form of governance in the world (Hamel and Keil 2015). For example, in the Indian context, civil society is represented by 'a relatively small section of people able to make claims as fully enfranchised citizens' (Roy 2015, p. 344). However, a vast majority still forms the so-called 'political society' based on their status as 'tenuously and ambiguously right-bearing citizens,' making a 'constellation of claims' (Roy 2015).[13] The majority is involved in development politics despite not being in the formal political procedure. In China, the mega-urban project is an outcome of these intertwined modalities—combining the state with 'planning centrality' and capital accumulation through 'market instruments,' succinctly known as state entrepreneurialism. The governance of large-scale urban development supports the statement that the state is more than a mediator. To an even greater extent, we regard the state's role as essential or central in governance modalities.

Reflecting on the mega-urban project of Lingang, how might it reveal the nature of state entrepreneurialism? Through the example of Lingang, we can see that Chinese new towns, as large-scale urban development in the suburbs, are examples of state entrepreneurialism. They are created by state-led city planning processes. They are not only master planned but also produced by planning policies, including, for example, the policy of land development quota, i.e., land consolidation in suburban areas. The new town is a planning product, for example, transit-oriented development (TOD), although the actual implementation may deviate from the original discourse of 'compact development.' The outcome may create considerable problems, such as long-distance commuting. Thus, one important task for new towns later is integrating industrial areas and living spaces. In contrast to piecemeal real estate projects, they are built under more strategic considerations such as economic restructuring and upgrading, innovation, and making global cities. The 'management committee' is a key mechanism for governance. The implementation requires market mechanisms, such as the financial approach to land development, and market agents, such as development corporations. We now elaborate on several aspects that constitute a state entrepreneurialism model.

First, the governance model is not an outcome of a growth coalition between the state and private-sector developers, even though the private sector participated in some projects, such as shopping malls and theme parks. In the early stages of development in the 1990s, real estate development stimulated urban sprawl in Shanghai. In China, many real estate projects driven by the property boom represent the coalition between the local government and real estate developers (Zhu 1999). A well-known example is Xintiandi—an urban regeneration project in Shanghai that relies on an overseas developer to regenerate old neighborhoods, leading to a total transformation of the area (He and Wu 2005; Yang and Chang 2007; Ren 2008). In the case of Lingang, the external input is also visible, as its master plan is designed by a global consultant firm—GMP based in Germany, but active in China. However, Lingang is not a project oriented toward suburban real estate development. It reflects some state intentions to upgrade Shanghai's economic structure into modern manufacturing industries. Although extracting land value is commonly used in China and in this large-scale development of Lingang, it has been intended for a different purpose. Lingang deviates from a finance-driven land-value capture project in many Western economies.

Second, it is not an entirely local project driven by the dynamics of the growth machine. It reflects a developmental strategy of the multi-scalar state (Li and Chiu 2020). Here, the notion of development is specific but can be variegated in different contexts, for example, environmental protection, ecological fix, socially harmonious development, and 'high-quality development,' meaning less environmentally damaging development with more economic innovation capacities like greenways and science cities.[14] At the top level, the project reflects the central government's intention for China to participate in the new international division of labor at a higher level as an innovative nation. This is seen as the expansion of Lingang into Lingang New District (*xinpianqu*), an extension of the science city. As a new town, Lingang originated from the municipal government initiative, assembling a development corporation—the Lingang Economic Group based on a successful Caohejing development corporation in central areas to fulfill its development strategy. However, district and township governments have also been mobilized to support large-scale development. The district government, initially Nanhui before it was annexed into Pudong, was in charge of the development of the town center. Pudong district later oversaw the whole Lingang new area. Initially peripheral to this development agenda, four town governments are responsible for the 'social management' of new residential areas.

Third, actual project management is complicated because of its large scale and the exceptionality of routine politics. It is argued that the nature of mega-urban projects leads to exceptions to traditional politics. However, according to the post-political literature, neoliberalism, or governing by market techniques, also advocates this. It has to assemble a wide range of actors and develop a network of horizontal coordination (Shen et al., 2020). Following the notion that the state is a set of networks rather than a unitary system (Cochrane 2007; Allen and Cochrane 2010), we have observed a series of governance innovations. The large-scale development involves the state and various developers, development corporations operating in a market form, and town and rural township governments. No single government can fulfill all the functions. The mission is also trans-scalar, often resorting to central government support, municipal government mission, district government administration, and town government management (Shen et al., 2020).[15] It contains more than one layer of government but multiple state actors at various scales. In other words, the administration of the mega-urban projects goes beyond a single government department or development authority (e.g., the 'management committee'). It shows the change from 'administration' to 'governance' in a complex network configuration. Despite this complexity, the state is central to the network relationship (Shen et al. 2020; Wu and Zhang 2022). There are collaborations between different levels of government and the government and 'local communities.' Recent governance innovation includes clarifying the division of duties between the mega-urban project and 'local communities,' called 'joint actions between the development zone and towns.' The development zone represents the mega-urban project through its management committee. The town government is responsible for social management, receiving a fiscal allocation. Hence, the coordination involves 'governance' rather than top-down commands, as their relationship is not simply the supervisory-subordinated one. Further, various development corporations representing their governments operate on the ground and collaborate with the local state and communities on shared market returns and risks (Shen et al. 2020).

Fourth, the mega-urban project is delivered through state development agencies—development corporations with the support of local town governments, with little involvement of the 'local communities' vision or its implementation. In other words, from a Western democratic perspective, the community voice is weak in urban politics because the politics or the political process is different from the Western oppositional political processes. It can be regarded as 'post-political' (see later in Chapter 3). Rather, the local administration unit presents community interests and participates in resource allocation. This has been a feature since the 'planned' economy in which resource negotiation is taken inside the administrative hierarchy. The state is paternalistic and supposed to look after society. The development process is largely arranged through professional planning. In this case, the rural community where the mega-urban project is located is weak in its administrative position. However, to implement the strategy smoothly, the administration of the mega-urban project, the so-called 'management committee' (the government authority responsible for the project), provided generous compensation to 'local communities.' The state maintains 'community engagement' through improving infrastructure and local living conditions, not because of the fear that these communities may resort to disruptive politics but rather out of the tradition of a paternalistic state.[16] Since the development strategy is aligned with the objective of economic prosperity, it is more receptive to compensation as a market transaction behavior. The redistributive effect (often known as the 'trickle-down effect') results from using 'market instruments' to fulfill its state strategies (Wu 2018b).[17] While in this suburban area, large residential neighborhoods are built in the form of 'gated communities,' their homeowners' associations do not create a base of private governance from which they switch to market provision or form 'local communities' in a political sense to negotiate with the government for service delivery in a wide range of residential communities (Wu 2022a). In reality, homeowners' associations or local villagers' committees are all instruments that help the government maintain social stability while looking after themselves.

In short, the large-scale project reflects state entrepreneurialism.

Modalities of mega-urban development

Lingang New Area reveals intertwined modalities of suburban governance in China (Wu and Shen 2015; Wu 2022b). These modalities include the state, capital accumulation, and authoritarian private governance, as shown in the governance of mega-urban projects and suburbia worldwide (Ekers et al. 2012).

First, the state plays a role in suburban development in neoliberal North America, post-socialist cities in Central and Eastern Europe, and developmental East Asia. In the US, classic post-war suburbs reflect the state's role in facilitating mortgage suburban homeownership and infrastructure development. The role of the state in Chinese suburban development is particularly visible and direct because the state not only takes the usual responsibility of the 'government' in its jurisdiction but also acts as a development agency (in the name of the 'management committee') in suburban development. It achieves this role through its subordinate development corporations.

Second, capital accumulation refers to the market dynamics and capital circuits. The investment in large-scale urban development projects, as capital inflows into the built environment, is intrinsically linked to the over-accumulation crisis of capitalism and the driving force of growth machines for property value appreciation. Here, Lingang reflects the expansion into a new space of accumulation in modern advanced manufacturing industries, a new space of high-tech sectors and science-technological innovations, as well as suburban development under 'land-based finance' through which the development of the real estate sector becomes pivotal in China's urbanization (Lin 2014).

Third, private governance reflects the role of the (civil) society. 'Governance' implies the 'devolution of responsibility from the state to both private sector actors and parts of civil society' (Ekers et al. 2012, p. 416; Swyngedouw 2009). The governance of suburban residential areas has been achieved through organizing local residential spaces into 'gated communities.' The government thus reduces some of the responsibility for services and maintenance and downloads them to these communities. The residential communities manage their properties through homeowners' associations, which appoint property management companies. However, in China, we do not witness the rise of corporate-style governance because of the ambiguous legal status of homeowners' associations. The traditional grassroots government, like 'street offices', town and township governments, and villagers and residents' committees, are still responsible for local administration and support the state's mission in this mega urban project.

Of these intertwined governance modalities, the mega-urban project manifests the centrality of the state as the governance does not evolve into a co-governance mode. Regarding the state–society relationship, the state leads the mega-urban project as a development mission, while local communities are mobilized and involved to support such a mission. At the same time, they receive the benefits of development. The state provides 'generous' compensation to local farmers to vacate the key development sites. The mega-urban project allocated land to the towns to develop their economic capacities. Later, when the new area was urbanized and the villagers converted into urban residents, the state provided public services through its administrative system of district, town, and township governments. The government attempts to integrate industrial development led by state-owned enterprises under the supervision of the Lingang Economic Group and urban life in the Lingang town center.

In short, the mega-urban project reflects 'inserting a corporation into the suburban area … to open up the suburban space for economic growth' (Wu 2022b, p. 194). This is beyond authoritarianism. To illustrate this point, we may ask: Who governs Lingang? Regarding this question, 'no single body can claim full control or is willing to take full responsibility' (p. 193). In this sense, a mode of 'governance' is emerging in China. However, this mode of governance is more accurately characterized as state entrepreneurialism because it is the particular way of intertwining governance modalities, which are more state-led, more for capital accumulation as a municipal-level strategy, and under-developed 'social innovation' and a weak voice of local 'communities.' The Lingang Management Committee, a special government authority leading the state mission, now faces the challenge of taking responsibility for comprehensive urban management. Still, there is no 'Lingang community' and no unified identity and interest of a suburban community based on residency (p. 194).

Critical reflection

This chapter began by reviewing the changing political economy in China to ground the notion of governance in its historical materialistic account. Following Chapter 1 on the concept of governance, we show the roles of capital accumulation and territorial politics in shaping the specific governance form, defined as 'state entrepreneurialism.' The term bears similarities to urban entrepreneurialism, but differs in terms of specific roles of the state. The state's role is embodied in territorial logic to stabilize capital accumulation and other intentions. Multiple objectives are raised to achieve such a logic. Economic growth is one of these objectives, as are policies to protect the environment and social reproduction. The concrete pathways to achieve these objectives are different in historical periods. These include institutional flexibility, scalar changes, and financial tools as an actually-existing form of state entrepreneurialism (Sun et al. 2023). As we have seen from the historical formation of state entrepreneurialism, the overall trend is toward greater state control. While the original notion of state entrepreneurialism stresses state 'strategies' and intentionality (Wu 2018b), practices on the ground might not achieve them. As seen during and post-pandemic, the instability might be caused by state actions, hoping to solve the tension between capital accumulation and territorial politics.

In this chapter, we provided a historical description of emerging Chinese state entrepreneurialism, which necessarily resorts to institutions unique to China, such as the legacies of state socialism, state landownership, and local politicians' career advancement based on economic performance. However, we also situate these institutions in conjunction with global capitalism and understand changing governance as the crisis management of China's capital accumulation.[18] The trait of entrepreneurialism in China emerged as institutional innovations to achieve the 'spatio-temporal fix' in its broad sense (Harvey 2003).

In the 1980s, the market mechanism was introduced to supplement the rigid state administrative allocation of resources. Decentralization and incentivizing local states led to rent-seeking and a political crisis in 1989. The re-opening of the economy to the world through export-oriented development led to inter-city competition and the reorientation of the local state to be 'business-friendly.' This is the most similar form to the typical urban entrepreneurialism depicted by Harvey (1989). Even so, we must not forget the 'intentional politics' or more reflexive state strategies, as seen in the remaking of Shanghai as a global city, partially representing a state strategy (Wu 2003a). As we see urban development politics in the context of global and national political economies, the state responded to a series of internal and external crises, which associated Chinese economic reform with global changes. From a 'state-theoretical perspective' (Jessop 2002b), the politics thus go beyond Chinese exceptionalism, as different economies share similar yet variegated state policies to historical moments and conjuncture.

This chapter provides a detailed example of Lingang's large-scale development. It shows that the relationship between municipal, district, and town governments and quasi-government development authorities (the management committee) evolved, leading to a governance form. Compared with the 'edge city' of business clusters, the case reveals the significant state rationality. This is illuminating, as an insight generated from China for other contexts (Robinson et al. 2021, 2022). The multiple objectives beyond financial consideration broadly resonate in quite different contexts. For example, in another mega-urban project, Old Oak and Park Royal in north-west London, the local state uses the specific institution of planning gain and 'viability test' to extract values to fund affordable housing targets and infrastructure (Robinson and Attuyer 2021). Conceptualized as trans-scalar politics, the mega-urban project demonstrates the role of the state in its specific circumstances of housing affordability crises and intensified local contestation. These studies also show the contradiction between these objectives, which means these state endeavors might not achieve their purpose as they wish, especially when the circumstance changes.

After the pandemic in 2023, China witnessed an economic downturn, a local debt crisis, low birth rates, and high youth unemployment rates. Large developers such as Evergrande and Country Garden face tremendous debt pressures and even bankruptcy. Some propose an austerity measure, like constraining the government's thrust for infrastructure investment, to rescue the economy while the state tries to steer it. Either way, the crisis exposes the internal tension between the state and entrepreneurialism and perhaps the end of state entrepreneurialism, as we have seen in the world-factory regime. The history of the global financial crisis tells us that the crisis did not end state entrepreneurialism but strengthened its role in financialization, turning it into a financialized state entrepreneurialism. That is, state entrepreneurialism further resorted to the debt toolkit. With a crisis both internally as shrinking social demand—a symptom of state infrastructure-driven capital accumulation—and externally as a changing global geopolitical order, will China escape it through a turn to the logic of capital accumulation and ending state entrepreneurialism? Perhaps it is too soon to give a definitive answer.

Notes

1 This section is derived from Wu (2023b).
2 For suburban development and post-suburbia, see Keil and Wu (2022), and also see the special issue in *Urban Geography* about the Asian perspective (Wu and Keil 2020). For inner city redevelopment, see He and Wu (2009) and Wu et al. (2022b), and for an overall picture of China's urban development, see Wu (2022a).
3 But the post-pandemic era may mean a greater financial constraint. Local states have to reemphasize the importance of private and foreign capital, because the government borrowing conduits are increasingly by the central government due to the concerns for financial risks and the downturn of real estate market which jeopardizes the dynamic of land finance. This point is suggested by Handuo Deng.
4 Thus, this is different from state profiting in real estate development, the original term of state entrepreneurialism used by Duckett (2001).
5 This interpretation reflects largely power decentralization, and deregulation, see Ong (2006), Roy and Ong (2011).
6 Other studies summarise the features of actually existing state entrepreneurialism, for example, politics of scale, regulatory flexibility, and financialization, see Sun et al. (2023).
7 Or, in the British government term, quango (Quasi-Autonomous Non-Governmental Organization). However, in China, these agencies act in a more corporativist way, as they possess financial capacities and assets. Compared with British urban development corporations with a limited number of staff, Chinese *chengtou* are more substantial market entities and have greater capital mobilization capacities.
8 In this book, 'strategic' state action differs from reactionary politics. The latter is more at the micro level of state and market relations. Here, strategic action could be a response to a more structural objective to maintain 'structured coherence.'
9 Here comes the tension within state entrepreneurialism. Some state actors use market instruments to achieve economic goals, even at the expense of more structural consideration. Not all actions follow the strategic goal, and they are often ad hoc, interfered with by local politics.
10 Some call this rural gentrification, see Du 2022, Yang and Loopmans 2023.
11 This paragraph is from Wu (2018b, p. 1389).
12 See also Wang and Wu (2019), Robinson et al. (2021), Robinson et al. (2022), Follmann et al. (2023).

13 For the original notion of the 'political society,' see Chatterjee (2004).
14 See Zhang et al. (2022) on the socio-ecological fix in Chengdu, and Wu et al. (2022b) about various state political mandates.
15 See Halbert and Rouanet (2014) for the concept of 'translocal territorial network' in financialization studies.
16 In this sense, the notion of 'negotiated authoritarianism' perhaps only describes the extreme case of social contestation and stability management.
17 See also Robinson et al. (2022) for comparison.
18 In many places, we tend to adopt a structural explanation of urban governance. That is, state entrepreneurialism as the evolving governance responds to historical and conjunctural crises. This echoes some 'reactionary' views of urban entrepreneurialism, dealing with exogenous and internal crises. China's evolving urban governance is part of a global process. However, the actions of the Chinese 'proactive' state differ from those of the proactive society elsewhere. Nevertheless, social development has progressed in post-reform China from an everyday perspective, even though these changes might not be present in formal politics.

3 Planning

State centrality and political mandates

Introduction

In the era of state socialism, city planning adopted a rationality different from the market rationale. It aimed to achieve resource allocation in the absence of a market. Compared with the regulatory planning rationale under the Keynesian welfare state, city planning under socialism presents a similarly strong role of the state in economic intervention but in a much more direct way of administrative command. It is also a more technical process within an administratively centered system with limited democratic politics. As such, planning under socialism was 'pre-political'—regarding the literature of 'post-political' planning, which will be explained later in this chapter, because it does not involve a societal-wide political process for its development and regulatory agenda. In short, planning was an administrative task for the state.

In this sense, China's planning system bore some characteristics of UK administrative discretionary planning based on planning permissions. Planning in China later introduced a zoning-like layer called the 'detailed control plan.' However, unlike US zoning and strong legal enforcement, planning in China relies much on the administrative system to enforce development plans. In China, the planning tradition shows a more proactive developmental style than rigid land use zoning in the US for certainty and security. In the US, 'zoning is not planning: it is a restricted instrument for districting' (Caves and Cullingworth 2024, p. 117).

After China abandoned direct administrative commands and introduced a more significant market mechanism into its economic governance, city planning has not been dismantled or streamlined. Planning has adapted very quickly to the new market environment. The influence of market-oriented reform over the Chinese planning system seems quite different from the detrimental impact on the planning system in Central and Eastern Europe (Hirt 2005, 2012; Andrusz et al. 1996) or the 'neoliberal attack' on its statutory planning in the UK (Lord and Tewdwr-Jones 2014).[1] Planning has survived market-oriented reform—as the 'phoenix rising from the ashes,' and its profession has been significantly expanded (Wu 2015b).

In contrast to politicians' hostile attitudes toward planning, Chinese city leaders regard planning as useful. The newly appointed mayor or party leader usually requests preparing a grand plan to demonstrate the vision of the new leadership. Because planning does not inherit the 'regulatory' legacy in market economies, it is not in a position opposite to the market. In the early era of 'growth-first' reform marked by marketization and globalization, planning became an instrument for promoting local economic growth. Nevertheless, planning is not undertaken on behalf of the market. As just mentioned, planning under socialism was part of the state apparatus. Its 'administrative' legacy means that planning is still for the state—for

DOI: 10.4324/9781003355175-4

local politicians to demonstrate their 'achievement in office.' In the earlier stage of economic reform, this achievement focused on GDP growth and hence was deemed less political.

As China moved further into state entrepreneurialism, as discussed in Chapter 2, planning shifted from local plan-making for the competitive city in an early reform era of market transition toward a more state-centered strategy under state entrepreneurialism. In 2018, the planning system consolidated multiple plans into a unified 'national territorial spatial plan.' Local governments made heavy use of the traditional urban land-use plan for place marketing and growth-oriented development. It has been absorbed into a more regulatory spatial plan.

Planning has been associated with more outstanding national political mandates (Wu et al. 2022b) (see Chapter 5 for the various policies related to urban redevelopment). It occupies a central position in national development and reflects state centrality in a broader sense. State centrality in planning means that the state prioritizes its objectives, including economic growth, in planning and dominates policymaking. However, achieving these objectives is more flexible during planning implementation. The means include market instruments and social mobilization in a later planning stage under the governance of 'co-production' with society. Increasingly, the national political mandate is no longer confined to the speed of economic growth but includes various objectives such as the Chinese dream and 'ecological civilization' (see Chapter 7 for environmental governance issues). As the Chinese economy has evolved into a 'new normal' of moderate growth, 'high-quality development' has been emphasized. Consequently, the mission of planning goes beyond its technical and economic objectives.

In the earlier stage of economic reform, the state strategy focused on economic growth. Accordingly, planning absorbs new market rationality through technocratic plan-making, inter-referencing—learning from elsewhere, and policy mobility. Large-scale urban demolition and environmental pollution led to social resistance and environmental crises, triggering development politics. However, the politics do not unfold through a growing self-governed (civil) society (Wu 2022a). The planning system is not open to a greater 'political' process.[2] Instead, planning strives to cope with these challenges and contain social conflicts while subject to the imperative of capital accumulation. The state must balance capital accumulation and the territorial objective of national modernization. In the earlier stage of reform, these logics were deemed more compatible with the 'planning for growth' process. Planning utilizes market instruments to achieve the territorial politics of national modernization and maintains state centrality (Wu 2018b). Now, there is increasing tension between these two objectives.

Planning centrality does not mean the state can always achieve its intentional goals. Often, planning implementation fails to deliver the targets. In planning, local governments usually present various discourses to demonstrate their alignment with the central government mandate but distort them in implementation for local interests. Therefore, the notion of centrality is not equivalent to a centralized state. It only suggests the rationality of the state, as a whole, over that of the market and society.

Centrality is not judged by the actual outcome of an effective state, despite recent administrative re-centralization. In the earlier stage of economic reform, the state acted in a market-friendly manner to realize its growth-first rationality. Consequently, it implemented a more decentralized state system as a 'weak' state regarding capital mobilization. When the economy experiences rapid growth and more associated problems, the state has an illusion that it can act more independently as it wishes—as if demonstrating a greater centrality. But this leads to more challenges and crises, forcing the state to collaborate more with the market and society toward the existence of an exaggerated version of a 'governance without government.'

The transformation of Chinese planning from pro-growth to more regulated can be seen in the top leader's vision for planning. When President Xi Jinping started his inspection of Beijing in 2014, he began his journey from the Beijing Planning Exhibition Hall. He visited an alleyway neighborhood (*hutong*) and expressed his concerns about the over-concentration of the population and economic activities in the capital. After his inspection, a subsequent strategy was formulated to relocate 'non-essential functions' outside Beijing to the capital city region comprising Beijing, Tianjin, and Hebei (the Jing-Jin-Ji region) (Figure 3.1). The city-regional plan thus becomes a state strategy. The large-scale city-region embodies the vision of the central government.

Other new planning experiments outside Beijing could be more locally initiated, like the micro-regeneration plan of the Yongqingfang area in Guangzhou (see Chapter 5 about regeneration) or the greenways and the 'park-city' in Chengdu, Sichuan province (see Chapter 7 about the environment), they all received the top leader's endorsement, reflecting new political mandates, such as 'building cities for people' and harmonious relation between human and nature under 'ecological civilization.'

These planning projects seem to demonstrate a post-growth mentality. In the West, the hegemony of 'growth' or other post-political strategies such as 'sustainability' creates a new post-growth condition for planning (Kaika et al. 2023). It is interesting to ask whether China's state entrepreneurialism produces a similar condition for its planning. Post-political planning

Figure 3.1 The new subcenter at Tongzhou in Beijing. The Beijing Municipal Government has been relocated to the new subcenter.

reveals the dominance of capital accumulation in the UK's planning process. As the history of Chinese planning reveals, planning for growth is part of national modernization. We might expect more visible territorial politics, including the post-growth shift.

Theoretical perspectives on neoliberal and post-political planning

In market economies, planning is justified by its role in reducing negative externalities, representing the 'police right' to enforce development control. The role of planning is achieved by containing illegal development to ensure public interests. Under the Keynesian welfare state, planning further takes a function of welfare delivery, for example, involvement in public housing development. However, this proactive function is somewhat limited compared to planning under socialism, as the market mechanism is still a foundation for economic development under Keynesianism. The state does not replace the market in market economies. Planning, as a state function, only corrects the market failure. Because of this regulatory tradition, planning is the opposite of the market and has its territorial rationality—resulting from democratic politics. It often works with the market but does not represent market rationality. Planning is associated with civic pride and better city design in the US. The City Beautiful movement is an important impetus for planning (Caves and Cullingworth 2024). Upscaled to a regional level, planning aims to solve transportation issues and protect natural and recreational resources (Caves and Cullingworth 2024). While planning seeks to regulate the market, private property rights are protected, and zoning is implemented within this legal framework.

After the rise of neoliberalism in the Thatcher era, UK planning has been subject to neoliberalization (Sager 2011). Planning is associated with public entrepreneurial activities, such as place promotion, marketing, and branding. The scope of planning has been dramatically expanded, with greater attention to the 'creative class,' culture-led regeneration, and new financing approaches. In the US, tax increment financing (TIF) opens up the mechanism to finance development through increased property value and, hence, the tax within a development district through planning (Weber 2010). In the UK, the Single Regeneration Budget (SRB) emphasized minimizing the development cost, and public–private partnerships (PPPs) introduced a more significant role of the private sector (Pike et al. 2019).

Starting in the 1980s, the direction of UK planning was based on a neoliberal agenda (Allmendinger 2016, p.7). Planning was regarded as the 'enemy of enterprise' and restructured—with new practices such as Enterprise Zones and Simplified Planning Zones where planning processes were streamlined. However, the initial attack on planning did not lead to the total abandonment of planning as the market requires some development certainty. Planning to curtail negative externality is still imperative for the market. As the machine for capital, the state apparatus serves the collective interests of the capital class. Hence, some functions to coordinate the market are indispensable (Jessop 2002a). Therefore, in the 1990s, there was a revival in planning. A 'plan-led' approach was introduced instead of an earlier deregulated and 'project-led' environment (Lord and Tewdwr-Jones 2014). The neoliberalization of planning, or neoliberal planning, reveals the processes of deregulation for the market and re-regulation to manage the market.

However, this plan-led approach was still market-based and growth-oriented. The UK's statutory planning system under the 2004 Planning and Compulsory Purchase Act produces spatial plans at different levels. Here, the discussion is mainly about the more strategic element of regional spatial strategies.

In the 2000s, the trend developed into a 'post-political planning' era. The notion of 'post-politics' stresses the fundamental nature of neoliberal governmentality as an instrument to close off the political struggles to support capital accumulation. It has 'replaced debate, disagreement, and dispenses with a series of governing technologies that fuse around consensus, agreement, accountancy metrics, and technocratic environmental management' (Swyngedouw 2009, p. 604).[3] However, this post-politics consensus does not mean the total disappearance of the political. It is a particular form of exclusionary politics. The 'properly political' resurfaces afterward in a non-democratic manner (Swyngedouw 2009). That is, depoliticization represents a shifting of the political, rather than its eradication (Allmendinger 2016, p. 147; also see Rancière 1999; Žižek 1999 for the original concept of the 'political'). That is, planning has seen rising political contestation.

The post-political condition regards planning based on democratic politics as a constraint on economic growth. New techno-managerial methods have been introduced to achieve consensus for a development agenda. In the UK, the introduction of spatial planning represents the post-political condition. It stresses partnership building and more collaborative approaches involving experts and local communities to deliver growth targets. The governance techniques include performance indicators, participation and consensus, fuzzy concepts, and soft spaces (Allmendinger and Haughton 2012).

This 'neoliberal spatial governance' uses consensus-building techniques but does not address the 'properly political'—here social inequality and environmental challenges—and maintain an existing police order (Allmendinger 2016; for post-politics and planning, see also Metzger 2017). The planning process, through the involvement of professional consultancies and social organizations, creates political activities, but in the end, it represses rather than encourages the possibility of change. The post-political condition marks the suppression of the 'political' by 'politics' (Mouffe 2005; see also Metzger 2017). Here, politics refers to the governance procedures and routines, while 'the political' means fundamental conflicts between different social groups (Mouffe 2005). Post-politics is also post-democratic and authoritarian (Fearn and Davoudi 2022).

Under post-political conditions, the previous regulatory land-use planning was transformed into spatial planning. For example, regional spatial strategies were formulated across administrative boundaries to promote collaboration between cities and form city-regions. However, these regional planning strategies were abolished. Since the 2010s, earlier enterprise zones with a simplified planning approach have been revived into 'local enterprise partnerships.' For major infrastructure projects, a new planning process is established at the national level, removing the influence of local resistance. The trend of spatial planning is arguably toward an authoritarian turn, 'with a shift of emphasis from techno-managerial to executive-punitive practices' (Fearn and Davoudi 2022, p. 347). The planning process is achieved through administrative management and deal-making (Gibson et al. 2023). The viability test imposes a condition to guarantee developers' profits (Foye 2022). Planning relies on seemingly incontestable concepts such as sustainable development and smart growth (Metzger 2017). As such, although planning promotes public involvement, it actively restrains the political influence of social groups, generating a superficial appearance of legitimacy (Metzger 2017 p. 185). Post-political planning leads to the foreclosure of public dissensus (Dikeç and Swyngedouw 2017).

However, recent debates over post-politics disagree on whether it represents an achieved condition or a contested trend. The notion of post-politics may reinforce the perception of 'there is no alternative' because 'labelling cities "post-political" risks treating depoliticization as a condition that has been realized, rather than a tendency that has been hold' (Davidson and Iveson 2015, p. 546). The process of consensus building does not manage to close off political contestation. The urban is a 'political space of resistance and emancipation' (Beveridge and Koch, 2017, p. 31). Instead of simply reflecting post-political urban conditions, depoliticization re-articulates the 'boundaries of political agency' and 'possibility in urban politics' (p. 40). For example, social resistance puts pressure on political parties and leads to changes in transport infrastructure projects, as shown in Melbourne (Legacy 2016). Some large planning development projects did not manage to go through the political process of planning because of public resistance.

A new social movement contesting austerity shows that the urban is becoming increasingly political (Uitermark and Nicholls 2014; Uitermark 2014; Beveridge and Featherstone 2021).[4] The politics is not limited to formal politics and procedures; it also includes insurgent citizens (Swyngedouw 2018). Recent anti-austerity social movements occurred outside formal politics through everyday contestation and struggles (Gerlofs 2019).

Although post-political planning suggests a shift from a political process of planning to a post-politics managerial approach, it must be noted that planning in Western democratic market economies has always been 'techno-managerial' (Metzger 2017, p. 189; see also Lord and Tewdwr-Jones 2018). While development politics involve a more liberal democratic process, planning might not be a political arena in the Keynesian welfare state. Economic decisions are made on a market basis, and the state intervenes only when the market fails. Politics are played in a broader political system to which planning is subjected.

The planning style in market economies is less direct and proactive than planning under state socialism in China. In Chinese state socialism, the city planning system was also technical and managerial. City planning was not the arena for decision-making regarding investment and development. The decision was made under the state economic planning system, involving formal and informal negotiation. Even in the reform period, while planning became more proactive, it mostly facilitated the local government to solicit and bargain resources from external investors and the central government (Wu 2015b).

In sum, the theoretical perspective on neoliberal and post-political planning in Western market economies provides insight into the impact of market hegemony on planning as a state political intervention. Rising neoliberalism creates a post-political condition where planning is streamlined and restructured toward fulfilling market development. Such a tendency is to eliminate growth obstacles or contain social contention through consensus building to achieve growth targets. Thus, planning is more oriented toward market development. Because of this transformation with neoliberal governmentality, the planning process constrains the 'properly political.' However, the issue of inequality, as a political issue, cannot be eliminated. It either resurfaces through more coercive and confrontational manners or is reinvigorated by everyday politics (Beveridge and Koch 2017). It might be appropriate to argue that planning in the UK cannot be fully characterized as post-political, as shown in the rejection of many housing development applications and the failure of development planning projects. The UK planning shows an anti-development rather than a pro-development feature.

This understanding of the consistent techno-managerial feature helps to bridge the gap between the theoretical perspective on 'post-political planning' and Chinese planning practices because the relevance of post-politics theory is its context of Western democratic political systems and thus has a limitation in China's state-centered system. China's planning might not be the negotiation and consensus-building process before and after economic reforms. But this does not mean that there is no development politics in China. The politics present differently within the state, involving central–local relations and negotiation between government departments and state 'work units.' The pressure from society is not manifested as a direct antagonistic relation between the state and society. Still, it is placed on different political factions of the party-state because the party internalizes disputes and negotiations within its system.

The introduction of a greater market mechanism in China has driven planning activities beyond its administrative role and the technical task of the government to interact with more actors in society and the market. After introducing market rationality, planning plays a greater role in interfacing the market and society with a seemingly more open planning process. In the 2000s, China witnessed a similar trend of 'planning for growth'—the discourse of growth is indisputable. Hence, it restrains the state's diverse interests and social contestation against market development. Technical experts are involved in an increasingly more plural planning process. Planning exhibitions are used to disseminate city plans. However, in the 2010s, the political nature of the planning became more visible when President Xi Jinping announced a Chinese dream. The technical rationality under 'planning for growth' does not eliminate the political nature of market-oriented development. Critically thinking about neoliberal and post-political planning, we now turn to the history of city planning in China.

A brief history of city planning in China

China has a long history of using cosmology and geomancy in the design of city buildings. However, its city planning as civic design and land use regulation appeared in the late imperial era in the 19th century. Modern infrastructure and construction methods, such as paved roads and building codes, were developed in the extraterritorial space of treaty port cities, such as the French Concession Area of Shanghai. These treaty port cities began to see the contrast between the under-serviced Chinese part and the European quarter built through modern city planning. Some major urban plans, such as the Greater Shanghai Plan in the era of Republican China, were prepared under Western influence. Through the professional exchange of architects and planners, city planning was a deliberate effort to learn from the modern West.

Trained in the West, Chinese planners proposed a grand modernist vision to modernize the nation. City planning is thus closely related to a national modernization project rather than a simple local land use plan. These grand city plans did not materialize because of constant warfare and unrest. From this history of early planning practice in China, we can see that planning was an effort to modernize traditional China regarding building codes, land use control, and population redistribution in the metropolitan region. However, the attempt to plan modernization failed for political conflicts.

This technical tradition of planning in China is revealing because planning represents a modern and rational approach to managing urban life. The plans largely adopted a blueprint style, hoping to create a new life by building the built environment. This modernization tradition is coupled with the tradition from the imperial period, specifying the norms and standards of city building. The origin of planning in imperial China reveals that Chinese planning has always been a political project beyond civic design. These two traditions continued to influence planning in the socialist period. On the one hand, planning should represent the principle of state socialism. On the other hand, it was primarily a technical task to assist industrial development.

In the socialist period, the planned economy was established. A powerful state characterized the system. The main objective was to modernize the nation through industrialization. Another related objective was to restrain urbanization, that is, to control the population inflow into cities. This is because the state wanted to concentrate its resources on industrial development. The consumption-oriented city was regarded as non-productive. New industrial towns were developed in the suburban areas. In 1959, Shanghai completed its master plan. The plan adopted the principle of population redistribution to industrial new towns across the metropolitan region. However, in reality, suburban towns were underdeveloped because of the lack of investment. Industrialization is primarily concentrated in the ring adjacent to the central city. Industrial development in inner suburban areas was a key feature of the socialist city.

Despite a planned economy, an intriguing question is to what extent the socialist state can plan its city comprehensively. The socialist city was divided by separated rural and urban areas, leading to a fundamental urban–rural dualism. Planning was for the city under the state domain. It was not for the rural areas. The rural area was largely outside the state domain of welfare provision, although it was subjected to the state-compulsory purchase of agricultural products. Rural areas were not under the control of city planning. The city was also quite fragmented. The state work units were asked to manage their land uses and functions within their boundaries. The coordination at the whole city or metropolitan scale was difficult and weak. The city government only intervened and regulated the residual space between state work units, such as the roads and public facilities constructed by the municipality. It was difficult to achieve comprehensive city planning.

Development control was absent in the socialist city. The process of city planning was not 'politicized.' Bargaining for resources was not achieved through city planning. The plan was not an outcome of negotiation between work units and central–local governments. City planning was only the last stage of the spatial arrangement of economic development plans. It was a technical task, assisting the 'materialization of national economic plans.' The economic plan set the economic development goals. City planning was not involved in development politics. The coordination of development was achieved through the negotiation and balance of economic sectors. State work units self-built their living quarters. The city planning process had limited influence over the internal land uses within these work units. The subdivision of land uses within work units was outside the planning sphere.

The investment mechanism for city-wide infrastructure was absent because infrastructure investment was regarded mainly as 'non-productive.' In addition to this ideological reason, the city government could not recover its investment in infrastructure. The investment did not generate a return and was a purely financial burden.

As a result, planning did not need to negotiate for 'public interest.' This contrasts with planning in market economies, which is justified based on 'public interests' rather than a logic of capital accumulation. Instead, the negotiation was achieved within the state system, for example, between central and local governments, economic sectors, and work units, rather than between the state and society. In other words, planning did not represent the society to negotiate its interests with the state.

Despite the non-political nature of planning processes, the socialist city was a political project. Its built environment is full of political symbolism, demonstrating 'socialist monumentalism,' with grand public buildings and squares to showcase socialism's superiority to capitalism, and the street was used for the political parade (Andrusz et al. 1996). The land use in the central areas was less oriented to commercial uses. With the absence of the market, the private realm was diminished. The socialist city forged a new form of collectivism based on state-organized collective consumption.

The political nature of the socialist city does not mean that the state could transform the built environment as it wished. Although grand buildings were constructed, the social city did not rebuild its city center. Many low-rise residential buildings in inner areas remained, often in a dilapidated state. In contrast to the newly developed industrial areas, the inner city lacked maintenance. For example, in Beijing, the traditional courtyard housing remains in the densely populated *hutong* areas (Figure 3.2).

In short, city planning was essentially a technical activity within local governments. It did not involve regulating behaviors outside the government, either as the market or society, because the market was absent. Society was managed through 'traditionalism,' where there was a high level of social integration between the state and society (or the state in society) (Wu 2022a). This traditionalism was based on the low percentage of the urban population and the self-contained cellular structure of work units and economic sectors.

The political nature of the socialist city began to diminish after the economic reform, and the state capacity declined initially. For example, market-driven real estate development relocated many residents to peripheral areas. The influx of rural migrants into low-quality hutong areas changed the social composition. Courtyard housing was extended to accommodate low-income residents. Temporary shelters were built within the courtyard, leading to the deterioration of housing conditions. Old housing areas were often redeveloped through a wholesale demolition approach or a new courtyard style with a changed social fabric.

In contrast to the 'depoliticizing' of city building and a more significant role of the market in economic development, planning, as a state function, was restored after the Cultural Revolution (1956–1976). The planning system evolved into a complex one—with multiple

Figure 3.2 Beijing's alleyway housing (*hutong*) area in inner cities and the financial district in Chaoyang district in the distance.

plans until 2018 when they were consolidated into one system. There are three main types. First, the five-year plan was under the National Development and Reform Commission (the original Planning Commission). Traditionally, the five-year plan was the economic plan, setting economic growth targets. Later, the plan evolved into a spatial plan. It provides a general spatial arrangement of national economic development that is more 'conceptual' and less detailed than land uses.

Second, the Ministry of Housing and Urban Rural Development implemented a mainstream urban plan. This plan comprised an urban master plan and a detailed construction plan. They were mainly urban development plans. Third, under the Ministry of Land and Resources, the land use plan was initially for rural areas but later extended to the whole national territory, mainly for controlling agricultural land.

The history of the Chinese city planning system indicates that it has been strengthened along with more market development. The urban planning system originated from industrial site planning after 1949. The 'city' plan was an expanded industrial plan with residential and other infrastructure components. The system lacked any legal basis. The first piece of legislation was the 1990 City Planning Act. This gave the city government power to perform development control. In 2005, city planning regulation was revised to include rural areas. Later, in 2008, it became the revised Urban and Countryside Planning Act. The rural area was formally incorporated. The Act came quite late because, at that time, the land use plan under the Ministry of Land and Resources had been well-developed with a mechanism of land control.

Based on the land use plan, the 'conventional' city plan was absorbed with other types into a consolidated 'national territorial spatial plan' in 2018. It became the single planning framework. From this history, it is clear that city planning evolved into a governance mechanism. The single planning framework is now under a newly established Ministry of Natural Resources. This significantly strengthens the function of land use regulation (instead of 'planning for growth').

Market reform has led to notable changes in the planning system, which allow the planning function to be more compatible with market initiatives and entrepreneurialism. First, planning became more 'pragmatic,' with less regard for political ideologies and an overall pragmatism promoted by Deng Xiaoping. Second, less ideologically oriented, achieving local development goals was important. The city government has a greater influence over the goal and the actual way of development. City planning reflects the vision of municipal government. Third, the development control mechanism strengthened the power of local states. In reality, it was difficult for the local government to control urban expansion because the impetus for expansion came from the government itself. Fourth, more layers were introduced into the planning system to strengthen the development control, making it more complicated. Even so, it was ineffective, as the growth-minded local government largely captured planning. This eventually led to its consolidation into a more control-oriented land use plan.

From this history, we can see the regulatory function of planning has been strengthened along with marketization. We will now elaborate on the consequent governance change after the injection of market rationality and how planning evolved into state governmentality.

The injection of market rationality

In market economies, planning is justified by its rationality to reduce or constrain the negative externality of market development. Zoning in the US strongly regulates development by separating land uses. Planning actions are based on zoning. Detailed physical controls over land use to prevent unwanted uses from invading desirable residential areas become the distinctive character of US land use planning (Caves and Cullingworth 2024, p. 71).

Planning represents a rationality different from the market. Hence, planning is, in essence, agonistic toward the market. It is based on the 'police right' to maintain public order by enforcing the 'development control' over land. The UK adopts a development control mechanism in a plan-based system, which requires all developments to apply for planning permission before making land use changes.

In a centrally planned economy, planning has a different rationale. Planning helped to inform the process of resource allocation. It assisted the government at various levels in allocating resources for national industrialization. In this sense, planning is part of modernization, as we have seen in the history of Chinese planning. In other words, planning aimed to regulate the behavior of governments. By this origin, Chinese planning was not 'hostile' to the market, while planning in the West aimed to regulate market behavior.

In the post-reform period, under globalization and marketization, planning is given a new mission to build a competitive city. That is, planning is for economic growth, specifically increasing GDP. Market rationality has been injected into planning, which can be regarded as a process similar to 'planning neoliberalization.'

Under the overall turn to neoliberalism, UK planning came under a 'neoliberal attack' (Lord and Tewdwr-Jones 2014). Planning was regarded as the 'enemy of enterprise' (Lord and Tewdwr-Jones 2014). The neoliberal ideology creates a hegemony of 'growth,' similar to what we have seen in China—'the growth is the truth' under Deng Xiaoping. Consequently, the comprehensive land use plan has been streamlined. The role of planning has been reduced, and a new style of planning was introduced, which turned the process toward 'project-led planning,' focusing on large projects instead of comprehensive land uses. Market contracts were used to enforce planning (Raco 2014). The deal-making between the public and private sectors replaced previous more democratic planning procedures, described as post-political planning, and, hence, this is known as post-political (Gibson et al. 2023).

However, planning also survived this neoliberal turn, as the market requires some certainty and the development order. Market development requires a planning system to deal with negative externalities. Nevertheless, planning only plays a limited role in ensuring the order of development and giving certainty to the development sector. The introduction of the viability test seriously undermined the claim for 'planning gains' in the UK planning regime (Foye 2022; Fearn and Davoudi 2022).

Planning in China also faces challenges from the market. But it has survived during the market transition. Practically, there is a great demand for planning. The development of world factories especially requires large-scale infrastructure provision and land development. Facing rapid urbanization to provide the primary conditions for social reproduction, old cities have been redeveloped, and new industrial zones and new towns have been constructed. This created a demand for planning, similar to state-led industrialization in the 1950s, which required planning to prepare factory layouts and residential quarters for industrial workers. Thus, planning is more than 'place marketing' or imagining by local governments. It provides a material condition for China's new industrialization under the global economy. Spatial production under globalization requires more rather than less planning.

Hence, there is a capital logic for planning, i.e., capital accumulation requires planning input. For 'land finance' and land value capture, planning plays an important role in delineating land parcels into tradable assets and collaterals with attached planning conditions and constraining the land supply and intensity for the local government to realize its monopolistic rent (see Chapter 4 on financialization).

In the West, capital accumulation, as the expansion of the capitalist market, is underneath the growth machine. Neoliberal planning practices under urban entrepreneurialism support the dynamics of the growth machine. Here, China has seen similar changes, especially in the earlier stage of economic reform. In addition, as part of the state apparatus, the planning system also supported a larger nation's political agenda, which was also more growth-oriented. Here, the capital logic for planning is not confined to the local land finance but also a logic for expanding the national and regional economies. In contrast to capital relocation within the US due to a weak local planning system, China has seen the strategic concentration of economies in the Pearl River Delta, Shanghai, and the Yangtze River Delta with related regional plans. At least, these plans facilitated such a trend.

There are three possible explanations for planning survival. First, planning has adapted to a new environment of marketization and absorbed various market mechanisms in plan-making. In other words, the planning system considers the market demand, such as the need for real estate development. It uses a more market-oriented method, such as consultancy and policy mobilities. Planning helps place marketing and manages to suit the operation of the market.

Second, planning is used to solve the problems created by market transition. The introduction of the land market led to rapid urban expansion of Chinese cities. Just like its counterpart in the liberal market economies, planning aims to tackle the negative externalities of the market.

Third, planning plays an essential role in facilitating economic growth and, in turn, legitimizes the power of the state. In essence, this is 'planning for growth.' Unlike the first point, planning for the market, this explanation stresses that planning embodies state strategies and reflects state centrality in urban governance.

To elaborate on the shift toward market operation, we see that plan-making has been commodified through market approaches. This does not suggest that planning is purely market-oriented and largely captured by the market logic. Planning is, after all, an activity of state apparatus. The state maintains its control over the plan-making process. However, a series of changes occurred in the process of planning.

Planning rationality has shifted from a technical and physical design to a declaration for economic development. The aim of planning has changed from 'serving the production and facilitating the living' under socialism, which means planning to provide a layout of industrial development and arrange residential needs during industrialization to place branding, enhance economic competitiveness, and serve investment needs. The planning style has transformed from a defined blueprint design to strategic scenarios and policy recommendations. In terms of the input of the planning system, under the planned economy, commands from the supervisory government departments and national planning standards were the sources for plan-making. In a market environment, planning needs to consider the vision of political elites and the advice from professionals and consultants. Consultants are recruited through public procurement to provide planning services.

The output of the planning system has become more complicated than two-tier master-detailed plans, comprising various non-statutory plans such as strategic plans and project-based designs. Some are just 'conceptual,' reflecting a development strategy, and others try to visualize the future landscape. The approach of plan-making, i.e., how a plan is made, has been re-oriented from an internal government task assigned to the institute of planning and design to a mix of external planning consultancy, design contests, and the local institute.

The functionality of planning has expanded from producing government internal documents and development guidelines and coordinating government departments in charge of different economic sectors to producing external statements, slogans, and policy recommendations, convincing external investors, coordinating lower tiers of government (for example, district governments and suburban counties), justifying the need to exceed national planning standards and for more land quotas from the upper-level government. Overall, planning has been shifted toward accommodating market demands and absorbing the market into the plan-making processes.

From the above discussion, we can see that planning adapts well to the new environment of market operation by using market approaches such as consultancies. Through market re-orientation, planning seems to become a more independent activity and is no longer a government internal task, away from direct government intervention. It seems to develop its own rationality—similar to a trend of 'post-political' conditions in advanced market economies. It pays more attention to the market than it did under state socialism and becomes an interface between the state and market forces.

Under the initial market orientation, some planning activities became consultancy services. But looking back, despite this market approach to planning, it remains a government activity. More often than not, it reflects the government's strategic intention. Perhaps it is not planning under urban entrepreneurialism, where market rationality prevails; rather, China's planning is under state entrepreneurialism. This distinction about state entrepreneurialism was discussed in Chapter 2.

In the post-reform era, planning has become more critical. The Chinese planning profession has experienced significant expansion. The system of professional accreditation has been established. The number of registered planners has significantly increased. Many cities have built large planning exhibition halls, like the Shanghai Planning Exhibition Hall in the central area near the Peoples' Square. It presents a distinctive building style near the municipal government of Shanghai. A high-tech screen now replaces the once-gigantic urban model (Wu 2015b). The development of exhibition halls demonstrates the rising status of city planning. They reveal the importance of planning and the political ambitions embodied in planning.

In this sense, they do not represent 'place promotion'—often symbolizing urban entrepreneurialism under the dominance of capital accumulation logic. The newly furbished exhibition hall now presents a gigantic slogan by President Xi Jinping: "People build the people's city." Along the Huangpu River, another museum has been built to commemorate President Xi's visit to the pedestrian path on the riverbank where the mandate was proposed (Figure 3.3). It symbolizes a new rationale for planning. Well before the refurbishment of Shanghai City Planning Hall, the Beijing Planning Exhibition Hall presented a model of Beijing Bay, which was a starting point for President Xi Jinping's inspection of Beijing in 2014. This latter hall presents a vision of the top leader to the public.

Because planning helps to generate land revenue, it is often regarded as part of the growth machine in a market economy (Logan and Molotch 1987). Similarly, Chinese planning contributes to land revenue generation. Local political leaders recognize its role, but this role of revenue generation is a political endeavor. For example, the district party secretary prepared strategic functional zones in the Wuchang district of Wuhan. The commissioned was given to the China Academy of Urban Planning and Design in Shanghai. Consultant planners seemingly became very important in plan-making, as shown from a post-politics perspective, reflecting a depoliticized nature.

Nevertheless, the actual 'chief planner' is the party secretary behind the whole project, who steers the plan-making processes. Experienced planners know this 'political' intervention very well, and hence regard it as very important to understand the intention of local leaders and then 'translate' the intention into actual development layout. Planning becomes very 'technical;' for

Figure 3.3 The Museum of the City for People to commemorate President Xi Jinping's visit to the public path along the Huangpu River.

example, calculating land revenue and plot ratios to achieve the target. In reality, the politics of plan development plays a very important role behind the technical work, although the politics are primarily found within the state machine.

Planning helps the local government to develop mega-projects and pays more attention to design features. This is a design-led planning turn. For example, in the city of Zhengzhou, a new financial district has been developed in Zhengdong new district. This development added 150 square kilometers to 133 square kilometers of the urban built-up area, which is larger than an existing built-up area, thereby indicating the scale of urban expansion. The new development doubled the development area of Zhengzhou. In 2001, a design competition was launched. The well-known Japanese architect Kisho Kurokawa won first prize for creating an iconic water landscape of an artificial lake. Surrounding the lake is a cluster of sixty high-rise towers; beside the lake is a convention and exhibition center. The new district presents a new generation of mega-urban projects driven by land development in China.

The project has become quite controversial because of its aggressive land acquisition and concerns over property vacancies. However, the new district developed quickly, becoming a more mature urban area, and office spaces were filled after a few years. Because of a property-led development approach, it is tempting to regard it as an example of post-political planning. However, the case also represents a more state intention, here in this specific locality and historical period, a growth-oriented planning approach.

The momentum of urban expansion in Zhengzhou has even gone beyond the new district and move toward building a larger city-region comprising Zhengzhou and the city of Kaifeng, a county of Zhongmou between them. This was again conceptualized through the Zhengbian Strategic Plan (Wu 2015b, p. 97). In the south, an entirely new district of Zhengzhou airport district has been established, hosting a gigantic Foxconn factory for iPhone production.

But to what extent has planning become an independent force, standing against or constraining the state for public interests? Or, alternatively, does it form a coalition with the state to guide the market and society? Planning is essentially a government function—funded by local governments. In the most commodified form, the local government is the client of planning products, let alone the informal and instructive relationship between planners and local officials. Although the planning authority recruits consultants and provides additional technical consideration, the planning outcome largely reflects the local government's view. The competition for foreign investment and preparing a strategic plan through commissioned consultancy to build this city region suggest the capital accumulation logic and the non-political nature of city building. However, a close look at mega-urban development reveals state intention and centrality—through its role in land acquisition, infrastructure provision, and strategic intervention.

Under the entrepreneurial state, a new type of plan emerged—the 'conceptual plan'—which was similar to the spatial plan in the UK. The plan reflects the local government's intention through the technical assistance of the planning profession. The first conceptual plan was made in Guangzhou to help fulfill its ambition to maintain its status as the economic center of Southern China. The plan abandoned the conventional land use plan, introducing fuzzy spatial concepts such as growth corridors and development axes, which resemble the development of spatial plans under post-political planning. The conceptual plan is non-statutory, illustrating some of the local government's development vision. Making plans is also an important 'public' event, a chance to place promotion. However, this place promotion is increasingly associated with state intentionality, which is still influenced by the logic of capital accumulation but often goes beyond its logic.

All mega-projects, such as new financial districts and towns, suggest strong economic growth intentions. Increasingly, however, we also see other major state projects such as the Xiong'an New Area and the Zhangjiang Science City of Shanghai, which reflect more of the state's concerns when facing territorial and geopolitical challenges (Figure 3.4). Xiong'an New Area is a good example of how technical rationality is displaced by more political considerations. Although these mega-projects present different ways to associate with the market forces, they remarkably reveal some continuity, namely, state centrality and intention, dating back to the semi-colonial era of the Republic of China, when planning was a project for national modernization.

Establishing the 'national territorial spatial plan' in 2018 has strengthened the central government's guidance over local plans. Since then, planning has also reflected the balance of interests between central and local governments. The advice from consultants is only selectively used. If the consultant is from overseas, it is often used more for imagining rather than substantial input. Local planning institutes are asked to adapt the planning concept into a final version for the government. If the consultant is from an upper- or national-level planning institute, it brings in its understanding of the national political consideration and the local 'client' interests. Considering the nature of planning institutes and the power relations between planners and the state, it is difficult to believe that planning could ever become an independent force as a check and balance against the power of the state. In practice, the state input is overwhelming in the outcome of planning. Planning was not an independent force

Figure 3.4 The new high-speed train station at Xiong'an New Area.

before, and it is not such a force after the 'market transition.' Planning reflects 'governmentality,' embodies state rationality, and uses governance techniques. Although various stakeholders might be involved in the process related to planning, planning does not reflect a market force or directly societal needs. It represents more of a state development vision and agenda.

Planning has adapted well to the market environment through a series of changes. It facilitates market development and fulfills the vision of the state. When the state priority is placed on GDP growth, city planning acts efficiently—its professional has dramatically expanded; through knowledge exchange and policy mobility with the West, it develops the skills to deal with the market development, representing the most 'entrepreneurial' arm of the state, through new public management of planning tasks—public policymaking becoming a more product of consultancy. Hence, it facilitates market development in China. Nevertheless, in terms of motivation, planning has not been aligned with market interests. Planning has been transformed during the market transition. But this does not mean that planning is for the market. 'Planning for growth' needs to be understood by the governance of 'state entrepreneurialism'—reflecting the state's intention to foster economic competitiveness.

Although Chinese planners often advocate upholding 'professional values,' they do not represent a third force. Once planning institutes were 'privatized' to become non-state-owned enterprises. Major planning institutes either reversed their ownership or established an institute more closely related to the government to take on the assigned task. Planning institutes are affiliated with or closely work with the government. Often, the planning institute represents the government in organizing planning activities involving external consultants.

Public participation is still very limited, although recently, the planning system initiated a new wave of 'neighborhood planners.' Government input, either formal or informal, is crucial for planning the agenda. 'Planning for growth' primarily reflects government intention, state strategies, and its vision for future development. Although a wide range of stakeholders, such as consultants, neighborhood organizations, and the general public, are involved in the planning process, planning has not become independent, lacking substantial input from society. Planning often represents an imagined market demand. It is not sufficiently politicized—it is more a politics based on the government.

Nevertheless, this does not mean planning is free of 'political' consideration. It is now influenced by new political mandates such as 'ecological civilization,' neighborhood participation, and rural revitalization (see Chapter 5 for redevelopment). The politics of planning is managed in a state-centered regime, not through the contestation of multiple stakeholders as in a market economy. Instead, planning is not as technical as it seems, reflecting its association with 'big politics' (the political) because the state sets the development agenda. Planning reflects state intentionality or territorial politics.

In sum, planning does not set the development agenda without political considerations. It does not present a similar 'post-political' tendency. Nevertheless, different from 'neoliberal' governmentality indicating the logic of capital accumulation and capital power, it reveals the governmentality of 'state entrepreneurialism'—state centrality and political mandates. Indeed, China's recent planning shift shows the state's return—as the state's political reaction toward market rationality. For example, planning the city of Xiong'an as a 'millennium project' in the capital region of Jing-Jin-Ji reflects the attempt to seek an alternative development model beyond land finance through anti-congestion in the capital region of Beijing and more balanced regional development (Figure 3.5).

Unpacking 'planning for growth': state centrality

Reflecting on the justification of planning for economic growth, Chinese planners unpacked the meaning of 'planning for growth.'[5] First, they suggest that there are variegated practices and plural motivations beyond promoting GDP growth rates. Although Chinese planning institutes need to obtain consultancy fees to make plans, they are affiliated with the various levels of government and take the plan-making task as a requirement. They are not independent market actors.

Second, city planning is situated in the environment in which China is embedded in the global production networks. Thus, the motivation for planning is not only driven by the local growth machine. Planning serves a practical need for developing a 'world factory.' It is more than place promotion or profit-making for the municipality.

Third, planning includes place promotion and strategic control, such as the 'ecological red line' and greater inter-city coordination. Planning reflects the development intention and spatial strategies, especially at the national and larger regional scales (Wu 2016a; Li et al. 2023a). Although planning occurs at the local level, it involves a wide range of actors across scales. It reflects the state's endeavor to achieve optimal spatial production for the nation or territorial governance.

The reflection suggests that it is necessary to understand planning practices and rationality 'holistically.' Individual cases may present local irregularities and the distortion of the planning function. However, a systematic view of the Chinese planning system highlights its

Figure 3.5 The city center of Xiong'an New Area in the Beijing capital region.

function of governmentality. Through adaptation, planning is injected with new development rationality and equipped with new governance techniques. While planning rationale is justified based on economic growth, what do we mean by 'planning for growth'?

First, a straightforward justification is that planning is helpful for economic growth. Planning helps the local government achieve its GDP growth target, set by the central government and distributed across the administrative units. The growth rate is a key performance indicator during officials' promotions. Note that this is a very simplified explanation. The actual carder evaluation is much more complicated regarding economic achievement and political loyalty.

This explanation rightly points to an increasing consideration of economic growth in the earlier stage of economic reform. However, the explanation is actually state-centered because it places the behavior of the state system—official promotion—at the center of analysis. The value of planning is justified by its usefulness to the state, not by the market or actual effect on growth or development outcomes. The state regards planning as a useful instrument for achieving its objective.

Hence, planning for growth is not planning for the market (Wu 2015b). Planning practices changed because the planning system was part of the state apparatus when the mission shifted from class struggles to economic development. Under this new mission, planning is not considered a market constraint. Instead, the mission requires planning to provide the necessary infrastructure for the market. Planning assisted 'resource allocation' under state socialism, and its 'de-political' nature as a state techno-bureaucrat tool meant that other actors did not deploy it to take an agnostic function against the market or the state. This non-political tradition means that planning is adaptable to facilitate the state mission, now the market expansion. Also, like planning in a market economy, Chinese planning is welcomed by the market for its supportive role in land development.

Second, a structural explanation traces the role of planning to capital accumulation. The thesis of capital accumulation explains the shift of capital into the built environment as the driving force of urban development (Harvey 1989). Surplus capital is generated during production, which produces a contradiction of the capitalist economy because the existence of surplus capital means the eventual mismatch of market demand. Surplus capital led to capitalist over-accumulation. The development of the built environment is a 'spatial fix' to this over-accumulation, which transfers surplus capital into the circuit of urban development.

How is this structural explanation relevant to China? As explained in Chapter 2 on the Chinese political economy, industrial development was the economic driving force under the Cold War. The market development in the post-reform period featured rapid urbanization and urban development, which opened the scope of capital accumulation. Further, participation in the global economy means a new growth opportunity for China. Planning facilitates this 'spatial' shift—from import-substitution industries to urbanization and from internal circulation to capital circulation globally.

Planning helps the local government mobilize capital, a Chinese way of financialization discussed in the next chapter, to fulfill infrastructure investment needs. This is achieved by creating planning conditions that turn rural land and former non-marketable land into land collateral. Planning also helps the state to monopolize the land supply and hence achieve land value capture from land development. Therefore, planning's role is indispensable in capital accumulation. The meaning of 'planning for growth' is related to capital mobilization, land capital generation, and the general logic of capital accumulation.

Third, a political explanation reveals the role of planning in development politics. Planning is built into the 'growth machine' with Chinese characteristics—'growth' under the multi-scalar state. Planning supports the interests of landowners and, in the Chinese case, the local state and various state work units that are de facto landowners. The growth machine thesis concerns the politics of land interests and actors associated with land value. It stresses the 'motivations' of these actors in urban development—their vested interests in pursuing the increase of land value. The thesis introduces the scope of land politics into understanding planning rationale, which further requires us to examine the land development process.

As discussed earlier, the notion is relevant to China's planning regarding the deep involvement in land regulation, which benefits the local state as landowners. From this perspective, planning for growth boosts land and property values. The explanation stresses land profits rather than economic growth, as economic growth is a means to achieve the 'end' of higher asset values. The growth machine rightly highlights the importance of land development in China and the role of planning in land development.

However, urban politics and planning as part of development politics may not be confined to land interests (Wu et al. 2022b). A deeper investigation into the actually existing politics of development and life in the cities highlights the politics of redistribution and politics involving the social reproduction of everyday life as a post-colonial explanation (Robinson et al. 2022). Moreover, politics are not just local in terms of the physically bounded space of cities. The Chinese state is a multi-scalar and relational one. While the local state might be more interested in land, the central state increasingly emphasizes a 'harmonious society' and better environmental quality. The latter reflects the territorial dimension of a sovereign nation.

Overall, the rationale of planning in China is justified by a growth instrument. It embodies state centrality rather than market rationality. As part of the state apparatus, the planning system is always associated with the political mandate (a territorial rationale). From its earlier practice under the semi-colonial era to promote China's modernization, to assist resource allocation for the industrial-centered growth strategy in the socialist period, to mobilize capital and provide infrastructure to develop a 'world factory' under globalization, and to realize a Chinese dream of cultural and 'ecological civilization,' planning has been justified beyond its technicality.

However, the process of planning is often more technical than political. Economic reform, particularly in the land market, introduces development politics into planning. Planning deals with the tension generated by market development and, hence, the politics governing the market. However, the planning rationale is justified by the motivation of land finance (capital logic) and 'big politics' such as national modernization, state legitimacy, and the Chinese dream (territorial logic). Planning has a narrowly defined objective of market-driven economic growth and a wider political mandate of national development. Even though planning is associated with national political mandates, plan-making has not been politicized. The state maintains the technicality of its plan-making process to achieve its goals.

National political mandates

China's urban planning system has seen a significant change in 2018. Multiple plans have been consolidated into a single system of 'spatial plan for national territory.' The notion of 'national territory' indicates that the new spatial plan strengthens the control over under-regulated land use changes in the previous period of entrepreneurial governance. The administrative responsibility of planning approval was also transferred from the Ministry of Housing and Urban and Rural Development to a newly created Ministry of Natural Resources, which has been responsible for allocating land use quota. The former used the overall development quota and planning permission, which did not effectively contain development under growth-oriented local governments.

There is a stronger emphasis on the regulatory role of planning. Beyond re-centralization and strengthening central government control, the entrepreneurial form of the planning system – and associated planning consultant industries – have been reduced. The functionality of planning is needed, but not in its growth-oriented form. Environmental protection is a key criterion for evaluating the performance of local officials. To some extent, this is a reaction toward the earlier planning practices captured by the local government, which engaged in entrepreneurial behavior for urban development. Large-scale land-driven urban development increased the local government debt and financial risks (see Chapter 4 for further discussion).

Since the 2010s, the trend has been planning for 'high-quality development.' Planning is for achieving economic competitiveness and enhancing the state's governance capacities over urban development. The trend includes upgrading economic structure and innovation, improving the quality of the environment, and 'ecological civilization.' Promoting economic innovation capacities and enhancing environmental quality are two major national political considerations for planning. They are placed alongside economic growth.

Two examples illustrate the new efforts to promote economic innovation and environmental quality, the driver for which originated from the challenges of the low-value-added world-factory model in the 2000s. The global financial crisis sped up the transformation towards 'high-quality development.' The first is Shanghai's Zhangjiang High-tech Park. It reflects the state policy to upgrade economic structures and foster 'indigenous innovation' (see Chapter 6 for innovation).

Planning for Zhangjiang provides a land use plan for orderly land development in the market economy. It designates a 'national indigenous innovation demonstration zone'—not as land use zoning but as a spatial governance technique within which state policies are implemented. The zone is a state space for innovation. In 2016, the park was further planned for Zhangjiang Science City. The plan aims to build a new city rather than a development zone.

The high-tech park thus allows a much higher development density and aims to create an urban atmosphere and greater integration with Shanghai. Creating a new science city hopes to stimulate agglomeration and interaction, i.e., achieving a clustering effect. However, the development of Zhangjiang is more than a spontaneous clustering of high-tech firms based on their production linkages. It reflects the state's strategy to upgrade the industrial structure and enhance national competitiveness. Through this policy-driven development of high-tech space, city planning contributes to the national innovation strategy and Shanghai's role in China's globalization and national development (Figure 3.6).

The second example is about the environment. China now pays more attention to green development and environmental quality (see Chapter 7 for environment). New green spaces are planned and developed under the 'ecological civilization' ethos. The development of greenways represents this effort. The greenway system started from the Pearl River Delta in

Figure 3.6 The Huangpu River at the new northern part of the Bund, and the Pudong New Area is at a distance in Shanghai.

2010 as a green infrastructure. Within a couple of years, a massive greenway system had been constructed (Chung et al. 2018).

The greenways aim to improve ecological quality by maintaining the green space and providing residents with environmental amenities and recreation spaces such as cycling. In Guangdong, greenways became a popular destination for leisure activities. Other cities soon imitated the greenway plan because, compared with large-scale new town development, greenways need only modest investment but demonstrate visible policy achievement. Unlike large city squares and parks, greenways are smaller, linear spaces for everyday use. Their development is less controversial and not resisted by residents.

The Greenways has been upgraded in Chengdu. An extensive greenway system has been planned and developed. The city of Chengdu hoped to develop the whole metropolitan region into a 'park city.' Instead of building parks inside the city, the urbanized areas are located within a great park—a green environment. The creation of green spaces is associated with the development of the new district of Tianfu in the southern peri-urban area of Chengdu. *The Guardian*, a UK newspaper, asks whether the park city of Chengdu imports the idea of the Garden City at Letchworth into China.[6] But this is unlikely a result of policy mobility. The concept of the Garden City is not mentioned in policy documents, unlike the earlier planning practices using 'eco-cities' and 'transit-oriented developments' (TODs).[7]

Instead, the notion of the park-city, i.e., converting the whole city into a park, was specifically associated with President Xi Jinping during his visit to Chengdu in 2018. However, in practice, the park-city integrates more market-oriented land development and green space creation (see Zhang et al., 2022 and Chapter 7). The greenways and park-city reflect that planning is more embedded into the political process outside the locality. Rather than a result of importing a planning concept from the West, the greenway plan reflects the strategies of a multi-scalar state. These strategies include, in the Pearl River Delta, a new nature-friendly recreation space for 'social happiness' envisioned by the provincial government and in Chengdu, the central government's ecological civilization and the municipal government's tactics to expand the development sphere into a metropolitan-wide region while aligning with the central government directives.

Critical reflections

Chinese city planning, as strictly defined as a profession of 'civic design' and land use management, plays a less regulatory role. Instead, it has been proactive as a tradition. The semi-colonial legacy shows that planning was an engineering approach to fostering national modernization. The legacy of the socialist planned economy confined it to a technical device for resource allocation in the absence of a market. Planning thus is more prescriptive. This tradition led to planning being picked up as an instrument for promoting growth in the post-reform era. In this context, planning becomes an interface between the state and the market to solicit investment.

Market-oriented economic reform has injected a market rationale into planning. The system of planning has been opened up. Although planning has never been properly 'political,' the history of Chinese planning shows that planning has been subject to a large political agenda, varying across different historical moments. However, market reform has made a substantial shift in planning. Regarding market development and globalization, Chinese planning is an important instrument for 'growth.' In this way, planning is subject to a similar 'post-political' condition—growth is overwhelming, as represented by Deng Xiaoping's pragmatism, that 'growth is the only truth.'

Building upon this pragmatism, planning has been seemingly less political—it detached from a political mission in history, either national prosperity or socialist industrialization. Based on this overall rationale, a consensus has been formed. In reality, city planning as a professional entered a period of rapid expansion. Its status has seemingly been raised. However, when dealing with politicians, professional planners feel their professional values cannot be fully respected. There has been a deep feeling of frustration. In other words, outside its profession, planning continues to be placed under a large political agenda in the name of GDP growth rates or land revenue generation in this specific historical moment. Planners have attempted to justify based on market rationality—for maximizing land profits—or territorial rationale of the state to maintain an orderly development. Critically examining the notion of 'growth' reveals it has been a state strategy. The strategy, however, may vary in different historical moments. Now, increasingly, city planning, as always, represents state centrality and political mandates.

Understanding the Chinese political economy of state entrepreneurialism can reveal the political nature. The role of planning needs to be understood in the context of the multi-scalar state. Under 'planning for growth,' the local government tries to use planning to unshackle regulatory constraints imposed by the central government to achieve its growth intention. In contrast, the central government uses planning to restore regulatory control over localities and maintain a spatial order. The upper-level governments use the approval of local plans to strengthen the authority of the state, in addition to cadre promotion and personnel management. Planning is thus a new governance technique. Planning is deliberately kept as a technical task to avoid ambiguity, negotiations, and debates within the government system. Through planning, the central state strives to exert a more decisive influence over the landowning local states to govern the development and achieve national political goals.

The trend of using technical governance to maintain state centrality has become more evident since the mid-2000s and has been significantly accelerated under President Xi Jinping. During the market transition, we see a rising government expenditure in public service provision, city planning, and smart governance. Planning extends the state apparatus into the market operation as a technical governance device. As such, planning is not for the market but for the state (Wu 2015b). Planning tries to remove the informality introduced by market development and impose a new formality guided by the state. City planning plays a significant role in demolishing informal urban villages and creating new towns and industrial zones.

Because planning lacks a mechanism of public participation, it represents the interests of the state—focusing on growth targets and the land revenue of local governments and governable spaces and political visions of the central government, rather than 'public interest' or the properly political in a democratic framework, as planning itself does not involve negotiation between stakeholders. Planning is subject to a political mandate rather than opening up a political process. The recent rise of interest in public participation and neighborhood planners represents an effort to inform and communicate government policies to society rather than a process that shapes public policies.

The chapter examines planning practices in China under market-oriented economic reform. In conclusion, we interrogate the rationality of planning regarding neoliberal and post-political planning and discuss the logic of capital accumulation versus the territorial dimension of a sovereign nation.

Since the introduction of the market mechanism, planning seems to follow the logic of capital accumulation—'planning for growth.' In addition to capital accumulation, the state strives to maintain 'structural coherence' – solving the problems created by capital accumulation as a condition for continuing growth. It acts according to its perception and may not consistently achieve its original purpose. The state action is not against capital accumulation. As a modernizing economic system in the world system of capitalism, its economy is also subject to the logic of capital accumulation. But out of its effort to maintain structural coherence for capital accumulation, the state develops its rationality —territory politics as a sovereign nation, seemingly independent from capital accumulation.

From the discussion of early planning practice in the late imperial and semi-colonial era, the territorial dimension reflects national modernization. This is the politics of the national state, which has been a prominent and overwhelming political force in China throughout its history. Therefore, planning presents a remarkable continuity following this territorial logic beyond adaptation to the market environment. Planning is always perceived as a means to realize a more prosperous nation, although concrete planning activities are confined to techno-bureaucratic activities. At some historical stage, the territorial logic geared planning toward prioritizing the speed of economic growth. In other historical moments, planning was re-oriented by other state priorities such as ecological civilization, economic development quality and innovation, people-centered urbanization, and migrant social integration.

The transformation of city planning is discussed in the periodization of China's urban governance (see Chapter 2). For example, the global financial crisis forced the Chinese central government to prioritize growth through a financial stimulus policy (see Chapter 4 for financialization). However, the policy became less effective as China entered a 'new normal' of moderate growth. Along with rising financial risks and volatility, the property bubble caused severe concern to the central government. The state tried to stress the development quality and constrain real estate speculation and large-scale urban demolition. The new policy aimed to stabilize economic development and acted against its cycles. One objective was to reduce the local government debt and associated financial risks. Consequently, we witness planning consolidation into a new spatial planning system for the national territory, with a strengthened regulatory role.

Reflecting on the nature of planning under state entrepreneurialism, we can see that planning has become an instrument for growth and adopts more market-oriented approaches. However, planning is increasingly used for other strategic development goals. That is, the meaning of 'growth' is not limited to GDP growth, but also includes national development. Planning is still strongly influenced by and embedded in the political process. The general public still has little influence over the development agenda. The impact of national policies has been increasing. Planning is now more influenced by national political considerations than in the 1990s and 2000s. In short, planning reveals a territorial dimension—as governmentality—to govern the subjects in national territory.

Finally, reflecting on neoliberal and post-political planning terms, they derive from their specific context of democratic market economies. Hence, it is challenging to apply directly to China. However, they are relevant because they present intertwined capital accumulation and territorial politics. This is relevant because market rationality has been injected into urban development and planning in post-reform China. Chinese planning practices reflect their long history of interacting with a changing 'globalizing' world, from the colonial era, the Cold War, neoliberalism and globalization, and a return to more intensified geopolitics. 'Planning for growth' in China is similar to post-politics, where growth is the dominant political drive.

However, Chinese planning practices simultaneously reveal the root of territorial politics. Neoliberal planning in China is represented more as state pragmatism—a 'state exception.' As such, it is a governmentality approach. Post-political is instead a tendency rather than an achieved condition. In a democratic setting, this might be due to the urban as an increasingly important area for agnostic politics and contestation. In China, this is represented more as an incomplete capital logic—regulated by a reflective party-state interacting with the crisis of capital accumulation and changing geopolitics as an outcome of a larger crisis of state capitalism.

Notes

1 There is an extensive literature on the transition of planning under neoliberalism in the UK, see Lord and Tewdwr-Jones (2014), Allmendinger and Haughton (2012), Allmendinger et al. (2016).
2 For the difference between politics and political, see Rancière (1999), and its application to China, He and Qian (2023).
3 For the post-politics literature, see also Etherington and Jones (2018), MacLeod (2011), Wilson and Swyngedouw (2014). For original contribution, see from Rancière (1999), Žižek (1999) for original contribution to this literature; Lord and Tewdwr-Jones (2018) for the implication for planning.
4 Also see a special issue in *Environment and Planning C* (Beveridge and Featherstone 2021).
5 This is a special issue of China planning journal. It is available online (https://www.upi-planning.org/Magazine/IssueMore.aspx?Y=2016&I=3).
6 See https://www.theguardian.com/housing-network/2014/dec/02/garden-cities-china-chengdu-letchworth.
7 Interestingly, the term was briefly used by the previous party secretary, Li Chuncheng, to prevent urban sprawl. However, Li was jailed later. His successors in Chengdu avoided using the term in official documents. The term park-city was thus deliberately invented as a Chinese concept attributed to President Xi Jinping.

4 Financialization

The long shadow of the state

Introduction

China has experienced large-scale urbanization. It is achieved through both informal urbanization financed by individual households seeking better jobs in the cities and formal urbanization through large-scale infrastructure development. Informal urbanization is driven by millions of rural migrants flowing into cities. The relocation of these migrants is primarily job-related. This creates a challenge for providing affordable housing. In practice, rural migrants live in urban villages, which are composed of informal private rental housing. The construction of informal housing is self-financed mainly by extending existing rural villages around the cities. Thus, cities expanded rapidly through a process of extended informal urbanization (Figure 4.1). Cities have also been developed through formal large-scale infrastructure projects, particularly mega-urban projects such as new towns, as discussed in Chapter 1.

How are large-scale urban development projects financed in China? To what extent have Chinese cities been financialized? Has the changing ways of financing infrastructure led to corresponding changes in governance? What is the implication for urban governance? Has land-based finance led to financialized urban governance? These are the questions to be addressed in this chapter.

This chapter shows that Chinese cities have initiated large-scale urban development and redevelopment programs. They need to raise capital to finance these developments. However, these developments are not financed solely by municipal public finance. Infrastructure-driven urbanization is financed by financial operations, especially through land collateral. Hence, a financial perspective is required in addition to understanding the fiscal aspect. We also need to understand the state's intentionality in financial expansion; namely, why the local state needs to raise capital from land operations. Is this for revenue generation or for the pursuit of other state strategies? The chapter highlights the issue of local government debt caused by land financial operations. As local governments finance development through bank borrowing, the question is: how can the government borrow? Where does the money come from? The chapter also discusses important changes which have occurred since the global financial crisis in 2008. The core concern of this chapter is how the practices of financing urban development have led to a process of financialization.

In the previous chapter, we discovered the role of planning in envisioning and delivering infrastructure development projects and, in turn, stimulating economic growth. China establishes development zones to attract investors and promote industrial development. Economic development generates taxes, and local governments invest in infrastructure development to fulfill a circuit. This is how the real economy develops. Chinese local governments sell the land cheaply to investors in order to promote industrial development.

DOI: 10.4324/9781003355175-5

Figure 4.1 The mixture of urban villages and mega-urban projects and extended urbanization in Guangzhou.

On the other hand, land constitutes a 'fictitious' economy, treating land itself as a commodity (Harvey 1982; Ghertner and Lake 2021). However, land value is an expectation for future profits. The government gains profits from land sales, and Chinese cities have seen rapid appreciation in the value of land assets. The cost of land acquisition increased. Land development, in terms of the production of landed properties, does not generate sufficient profits or directly contribute to public finance, which is widely known in China as 'land finance.' Instead, land development becomes, in itself, a source of development finance. Therefore, the significance of land is greater than its contribution to local public finance; rather, it is its leverage to raise finance. City planning supported the operation of land development. In addition, local financing platforms, that is, various urban development and investment corporations, have managed to gain access to the capital market. In this process, land becomes fictitious, as it is not a commodity *per se* but, rather, bears the whole process of capital circulation. However, the financial operation of land development is not entirely clear. To what extent does it reflect financialization?

The concept of financialization

The term 'financialization' is a relatively compounded and fragmented one (Christophers 2015; Rutland 2010). The extant literature interprets financialization according to three distinct aspects: as a regime of accumulation; as the rise of shareholder value; and as a change in everyday life (Rutland 2010). First, financialization is a new regime of accumulation in which a financial hegemony is developed. According to Krippner (2005, p. 174), financialization is 'a pattern of accumulation in which profit-making occurs increasingly through financial channels rather than trade and commodity production.' The dimension of the definition refers to the financial sector.

Second, financialization is concerned with the growing prevalence of financial logic in economic governance, for example, shareholder or bondholder value (Peck and Whiteside 2016). It is not only about profit-making through financial means, as suggested in the first aspect of the rising financial sector, but also about deploying financial instruments and imposing a financial logic. This second aspect of financialization indicates the intentionality of imposing the financial logic. However, the underlying intentionality is subject to debate when the state uses financial means. Seeing financialization as an imposed financial discipline—a financial logic—might be too narrow an approach, as financialization can be a statecraft that aims to maintain state power or simply 'stay in office' in capitalist economies (Pike 2023).

Third, financialization is seen as expanding the financial spheres even further into everyday lives, wider than in economic governance, for example, with social implications for credit and debt and a calculative approach to social life (Kaika and Ruggiero 2016).

As such, the meaning of financialization remains debatable (Christophers 2015; Ioannou and Wójcik 2019). Perhaps the most comprehensive definition is given as the 'increasing dominance of financial actors, markets, practices, and measurements, and narratives, at various scales, resulting in a structural transformation of economies, firms (including financial institutions), state and households' (Aalbers, 2017, p. 3).

Another definition sees it as the 'increasing role of financial motives, financial markets, financial actors and financial institutions' (Epstein 2005, p. 3). Although a rising financial sector in the economy indicates a growth in financialization, a stricter definition includes the specific intention to follow a financial logic. The former, however, refers to the widespread use of financial tools and the expansion of financial processes in the economy without necessarily following a financial logic, as was seen in the odd case of China. The latter refers to a governance shift. It is important to distinguish 'financing' from 'financialization' here, as the former refers to the use of financial sources. In contrast, the latter implies a consequential governance turn triggered by using financial means to finance development appropriately in the sense of governance change (Wu, 2023a, p. 38). This can be seen in Peck and Whiteside (2016), as they focus on the notion of financializing rather than financing, illustrating that financialization is actually a change in governance. They specify that it is a 'financial turn,' as discussed in Chapter 1.

Thus, the definition of financialization is relevant to a governance change, as argued by Peck and Whiteside (2016):

> Financialization is taken to refer to a historic process of systematic financial intensification which is reflected, inter alia, in an increased reliance on (and resort to) financial intermediation and financial engineering, along with a host of finial logics, metrics, and rationalities; in the empowerment of financial sector institutions and agents, including credit rating agencies, technocratic managers and overseers, bond market players, and legal advocates and arbitrators; and in the disciplinary roles played by shareholder-value pressures, capital-market interests, and the 'permanent economic tribunal' (Foucault 2008, 247) that is sustained budgetary restraint .
>
> (p. 237)

As quoted above, financialization is also seen as a governmentality method, such as using the technique in finance and making people value following the 'economic tribunal.' This definition is more stringent, as it requires a governance change to be qualified as a process of financialization, especially an increasing influence of the financial logic. The capital or financial logic is always intertwined with territorial politics, as has already been discussed in Chapter 1. The dominance of financial logic is often a matter of degree rather than a black-and-white distinction. However, as can be seen in the case of Chinese financialization, the strong role of the state is often a salient feature.

While all studies highlight the importance of finance, state–finance relations are variegated, ranging from finance limiting the state's capacity to the state facilitating financial operations and also furthering its own interests through financial means (Aalbers 2023, p. 796). From this state–finance symbiosis, the Chinese case seems to suggest that the state creates the condition for financialization and brings finance inside the state (Aalbers 2023).

The literature on financialization suggests two major dimensions to understanding a significant shift in the economy and polity. First, financialization is a rising sector in the economy and involves the widespread application of financial instruments. This criterion measures whether urban development has increasingly resorted to financial approaches. Here, a stricter criterion is concerned with a financialized way of economic development in addition to the usual development and consumer finance, for example, development loans and housing mortgages. Specifically, in the case of housing and often in the US context, it only refers to 'mortgage securitization' (Gotham 2009; Aalbers 2020). It is important here to distinguish development finance and financialization as the latter in the literature often refers only to the complex financial methods to extract values. Even the creation of a government bond might not be considered a financialization process in this case. Institutional investors in the rental housing market treat housing as a financial asset for rent extraction (Fields and Uffer 2016). More broadly, it refers to 'assetization,' for example, converting land or landed properties into an asset that can be treated in financial terms (Ward and Swyngedouw 2018). According to this understanding, a bank-based system under the strong influence of the state does not qualify as being classed as financialization, if securitization is limited (Shi 2023; Jiang and Waley 2022).

Second, financialization reflects the rising dominance of the financial logic in urban governance. The term financial logic is poorly defined as it mainly refers to maximizing value extraction. Strictly speaking, it is about a set of market exchange rules. This second dimension is translated into whether governance is financialized or a question about the 'financialization of the state,' that is, to what extent does the state's governance follow the financial logic—the logic of capital accumulation? The second dimension is about the transformation of governance.

As introduced in Chapter 1 about governance, in late capitalism, illustrated by examples from Detroit and Atlantic City, the New Jersey casino capital, there has been an arguable shift from entrepreneurialism to financialized governance (Peck 2017a, b; Peck and Whiteside 2016). Urban governance is characterized by a new form of dominance: austerity urbanism and debt-machine dynamics. Other cases taken from European countries confirm the first dimension, but suggest a more active role for municipalities that employ financial approaches (Ashton et al. 2016; Van Loon et al. 2019). Even in the US, financialization could reflect a more purposeful application of a tax instrument, for example, tax increment finance (TIF) (Weber 2010), which confirms the first dimension and even suggests a rising calculative approach, similar to the example of 'shareholder value' in financial operations (Peck and Whiteside 2016).

Financial applications like TIF indicate, therefore, the financialization of the state. Nevertheless, applying TIF does not mean that its intentionality is for financialization. This is also evident in the UK because the government does not treat public land as a financial asset; rather, it uses it to subsidize the operation of housing development (Christophers 2017). However, the outcome is still financialization as the private sector uses land in a financialized way. That is, the state action permits or enables financialization. More than enabling, the state may be directly involved in financialization by creating its own agencies for financing urban development, like urban development and investment corporations (*chengtou*) in China (Feng et al. 2022a).

Overall, the second dimension reflects an apparent financial logic in governance, which is generally confirmed by the literature on financialization in Western economies, albeit to a different extent. As seen from China's urban governance, fiscal expansion has led to the need to refinance debts. This seems to suggest that financial logic is in operation in this country. However, as discussed in Chapter 1, the state also represents territorial politics. In China, the protean state strives to cope with various crises and problems created by its own operations in finance. Hence, financial and territorial logic are intertwined in 'financialization' and 'de-financialization.' As we have seen, the term financialization is complex enough, let alone another related notion of de-financialization.

Here, we adopt a comprehensive definition of financialization, as introduced at the beginning of this section, as three interrelated processes of financial importance: the macro regime of capital accumulation; the meso level of firms in finance; and micro-level everyday financial interference. De-financialization can be viewed a process that reverses these trends. For example, financial rationality (logic) is restricted in financial operations, even under conditions of financial expansion. As such, the process of financialization is compounded and is not unidirectional, which may include, first, financial operation and, second, the financial logic behind the operation and all related actions. The Chinese case confirms the first, but does not fully confirm the second criterion (Li et al. 2023b). Here, we see the tension between these two logics.

Because of this tension, and the changing relationship between them, a question arises about China's financialization, as its 'political economy highlights the nature of Chinese state entrepreneurialism as developmental and managerial rather than financialized' (Jiang and Waley 2022, p. 1275). Given the limited securitized form of finance, the state principally uses the land to raise finance for development. Borrowing by local authorities relies on LGFVs and does not involve extensive private-sector finance. Therefore, as argued by state entrepreneurialism, 'what we see in China is not a financialized capitalism, but a growth strategy in which the state corporatizes itself to attain developmental goals' (Jiang and Waley 2022, p. 1279).

Similarly, examining the securities exchanges, it is found that the state facilitates financialization and exerts control over and manages financialization (Petry 2020). In the debate over financialization, the Chinese case highlights the limit of financialization due to its unique state–finance relationship. On the one hand, we see an increasing use of financial instruments enabled by the state, not only through deregulation but also according to public ownership (Whiteside, 2023). It even directly absorbs the financial operation into the state (Wang 2015). On the other hand, closer scrutiny suggests only a partial change in governance, which illustrates that the financial logic is subject to the state as territorialized social relations. The implication for governance is that financialization does not mean a neoliberal policy (as state-enabling financialization) but reflects a continuation of state-centered operation (Petry 2020; Jiang and Waley 2022), as highlighted in state entrepreneurialism (Wu 2023a).

From distinguishing financial expansion and financialized governance, it is appropriate to understand financialization as statecraft (Pike et al. 2019). The role of the state in financialization is extensively observed (Belotti and Arbaci 2021; Yeşilbağ 2020; Zhang 2020), and the state itself is financialized (Dagdeviren and Karwowski 2022; Peck and Whiteside 2016). The fiscal pressure, especially under conditions of austerity, pushes local governments to resort to financial instruments and operations in an attempt to generate revenue (Ashton et al. 2016; Christophers 2019). In this sense, it is 'state-led financialization' (Yrigoy 2018). Nevertheless, the state also constrains the process of financialization, which is known as 'de-financialization.'

De-financialization means that the state tries to constrain the financial logic and the operation of the finance sector. The term describes the restriction of financial elements in economic development (Li et al. 2023b). As with the all-encompassing term of financialization (Christophers 2015), de-financialization does not mean a decreasing finance sector (Li et al. 2023b). The state may encourage financial expansion while constraining financial logic. Social movements may challenge financial motivation and money-making through financial means and question the legitimacy of state policies, in turn forcing the state to restrict financialization (Fields 2018; Wijburg 2021; Karwowski 2019).

Hence, the statecraft may include elements of both financialization and de-financialization. The state may restrain the financial logic and internalize the financial operation into the state (Dagdeviren and Karwowski 2022; Ban and Bohle 2021; Wang 2020). As shown in the Chinese context, the financial logic is subject to the territorial and political logic of the multi-scalar state. It represents pursuing policy goals with financial instruments (Lake 2016; Li et al. 2023b). In China, recent policies aim to regulate shadow banking practices and introduce more regulated forms, such as local government bonds (LGBs), to finance urban development and incorporate them in the budgetary process of public finance (Li et al. 2022). It represents the selective use of financial tools and a de-financialization process through statecraft in financial governance.

Financialization and de-financialization are all possible policy options of the state. In governing urban development, some aspects of the financial operation are encouraged, such as the issuing of bonds. By contrast, other aspects are restrained, such as shadow banking, which is now subject to the fiscal budgetary process. The option is related to the *intentionality* of the state. The financial landscape in the post-pandemic period in China is characterized by both an expansion of finance operations and re-regulation through the multi-scalar state. The state balances financial logic and policy objectives (Li et al. 2023b; Pike et al. 2019; Kay and Tapp 2022). The Chinese case also reveals the multi-scalar nature of the state. Financialization began with the central government's stimulus package, and the central government also initiated 'de-financialization'—namely containing the financial expansion and logic—to manage the financial risks.

The process of financialization in China

The history of financing urban development

Chapter 2 introduced the subject of changing urban governance in China and the history of entrepreneurial statecraft. It suggested that this entrepreneurial statecraft has been financialized since 2008. That is, governance increasingly employs a financial approach. This section elaborates on the specific process of financialization. Although the local state has been using land value capture to finance urban development for a long time under the world-factory model, known as 'land finance' (which is elaborated later in this chapter), the process of financialization began with the global financial crisis in 2008. This section aims to explain the shift from land finance to land financialization (Wu 2022b), explaining how financialization, as financialized statecraft, emerges from the practices of financing urban development. Here, we provide a more historical and structural account.

China's financialization began with the management of the crisis of capital accumulation. More specifically, this was a debt crisis resulting from fiscal expansion in the aftermath of the global financial crisis. However, before 2008, China's world-factory model had also faced difficulties. The rising labor costs before the global financial crisis, due to a more protective social policy for migrant workers and the failure to provide the necessary conditions for social reproduction, led to the relocation of manufacturing industries to Southeast Asian countries. Just before 2008, manufacturing industries in the core regions of the world-factory regime, such as Kunshan in the Yangtze River Delta, faced difficulties and sought an increase in urban-based development.

The global financial crisis disrupted the capital accumulation of China's world-factory regime and led to massive unemployment among the country's migrant workers (Chan 2010). The Chinese government launched a 4 trillion Yuan stimulus package in response to the global financial crisis. Although presented as state investment in infrastructure, the package did not adopt a Keynesian model. Instead, it resorted to a financialization approach, allowing local governments to borrow capital through so-called 'local government financial vehicles' (LGFVs) (Liao 2015; Feng et al., 2022a, 2022b, 2023). The business model thus has shifted from finding external capital—an 'entrepreneurial model'—to financing urban development through financialization, as discussed in Chapter 1. Here, the subtle distinction between financing urban development using financial approaches and 'urban financialization' is clarified.

Financing urban development does not necessarily mean financialization. It may not require any resort to complex financial operations or capital accumulation rationality. It can, for example, be a routine operation for development companies to borrow capital from banks in order to finance their projects. The borrowing is often required to use landed properties as collateral. Land collateral is a specific form of using existing property ownership to finance future land development. The land system in China has evolved through several stages.[1]

Although the initial stages did not adopt a precise form of financialization, they did lay down the foundation of the system because they revealed the economic value of the land. In the 1980s, collective farmland was transferred into a household contract system. The rural village land was used for township and village enterprises in rural industrialization. The land thus had an exchange value. In the 1990s, local governments used land to attract external investors, and began to use land for development (Wu et al. 2007; Tao et al. 2010). In the 2000s, after housing reform and rapid rural–urban migration driven by export-oriented industries, China started a new phase of land finance. Given the municipal control over the land supply, the local government extracted profits from land sales, leading to a property boom (Lin and Yi 2011; Tao et al. 2010; Wu 2015b; Zhao et al. 2017).

China's land value capture can be dated back to an early stage of land reform. Immediately after establishing a land leasing system in 1987, Shanghai experimented with financing urban development through so-called 'virtual capital circulation' in the development of Pudong New District. In the absence of a land market, it was difficult to estimate accurate values (Wu 2000a; Wu 2023a). The development corporations in Pudong New District thus could not raise capital through land collateral. At this point the state credit backed up the land value—Shanghai municipal government deposited a fiscal budget-backed 'cheque' to the bank, indicating that it was willing to buy back the land if the development failed. With this specific public finance underwriting, the four development corporations managed to secure loans and initiated land development.

The profit generated from land development was used to pay back the loan. This is known as land value capture or land finance. Although land value capture can be seen as a process of financialization, it is different from the financialized governance approach, as a stricter definition of financialized governance requires a change in governance rationality. Land value capture can be simply used to finance urban development. In short, land finance and housing development through the real estate market still represent an approach of commodification.

The new stage of urban financialization is more structural in the sense of financialized state entrepreneurialism. After the global financial crisis in 2008, the linkage between land and finance was strengthened. The land is used as leverage for investment rather than 'land value capture' to generate revenue. In other words, the financial rationality has been strengthened—for the purposes of capital expansion. Land development does not simply aim to generate an income to contribute to public finance. Instead, it acts in capital mobilization to maintain the 'structural coherence' to cope with its crisis. The process of financialization has turned land into fictitious capital, and land development has become more speculative. The evidence for this financialized landed development is that local governments pursued land acquisition even though the current land development business was less profitable (Wu 2022b).

The cost of land acquisition significantly increased due to social resistance and the reinforcement of compensation standards. Local governments tended to compensate landowners generously so as to avoid large social conflicts and maintain stability. Given the huge costs associated with land acquisition, the motivations of land development are less associated with land profits as a contribution to local fiscal budgetary deficits, as suggested by the literature on 'land finance' (Tao et al. 2010).[2] Instead, land acquisition is just one aspect of state operations to cope with the financial crisis. While not excluding profit-making motivation, state entrepreneurialism offers a more structural explanation, as seen in the following discussion of statecraft in initiating, steering, and managing financialization.

Financialization

In the first stage of financialization, state-owned development corporations—urban development and investment corporations (UDICs) were into financial vehicles. Through government implicit guarantees and making land assets into collaterals, UDICs acted as more than developers. They created financial conduits in response to credit expansion under the stimulus package. One type of UDIC was the 'land reserve center.' The function of this organization focuses entirely on land banking for capital mobilization (Feng et al. 2023). The land reserve center represented the municipal government in acquiring land. It was a quasi-development authority, but one which could also take out bank loans with government guarantees as an enterprise status. The land reserve center was thus essentially an asset management company with privileged government status. The asset injected by the local government is known as 'reserve land.' The reserved land is often virtual because property rights have not been created or obtained, but the government managed to use the land to raise capital. The proportion of capital raised by reserved land to the capital backed with land collateral has risen sharply, suggesting that the function of reserved land was for capital mobilization rather than land utilization. The reserve land is, in fact, a form of local government guarantees—the right to its future proceeds rather than actual land ownership because the land does not have salable property rights. The financial risk is that the capital raised from the land is not for the development need but rather being used purely as a capital mobilization method which is to be paid by future land value increase. Because the reserved land has not been converted into land for sale, and even the plan for such a conversion to sell the land in the market is still uncertain, there still remains a tremendous financial risk.

Various state-owned enterprises raise capital through land and act as LGFVs, even if they are not actually development corporations specializing in land development. Following the global financial crisis, land has not been seen as a major means of attracting manufacturing investment due to the rising cost of development and declining demand. Local government injected land as an asset into UDICs in order to raise capital for infrastructure-driven urban development. UDICs are not necessarily LGFVs, however; before this late stage of financialization, they were simply state-owned developers. They might not actually carry out fundraising, although capital mobilization is usually needed to finance their development projects.

What has changed at this point is that in the late stage of financialization, some UDICs mainly act as fund-raising agents, using their development projects or assets to raise funds, which is not viable in the operation of development projects. This process of 'virtualization,' as described by Chinese researchers, produces a fictitious financial asset. This is similar to the assetization carried out by private developers in the UK (Ward and Swyngedouw, 2018). The difference is that in the case of China, this is achieved through state-owned enterprises or government agencies.

Four years after the fiscal stimulus package, heavy borrowing by development corporations and local governments put them under severe great financial pressure. The capital raised was in the form of short-term loans, which were used to finance long-term urban development projects, and they now required refinancing. This led to the development of a shadow banking system, a de facto financial liberalization that goes beyond standard financial projects under bank regulation.

Shadow banking is not illegal; it is simply banking off balance sheets. This included various forms of financial products. Wealth management products (WMPs) are rudimentary vehicles which target ordinary households. They offered a higher interest rate than savings and served as a conduit for various UDIC projects to access capital. They are opaque because they are off-balance sheet and do not contain payback guarantees. This product experienced faster growth after 2012. Another form of product is the trust. Through collaborating with trust companies, municipal governments obtained capital for their projects. Trust companies created non-banking debts for the local government (Chen et al. 2020). The overall trend has shifted the debts to off-balance sheets by converting earlier bank loans to WMPs and entrusted loans.

Non-standard banking is also a result of tightening financial control over debt. Because of continuing financial pressure and the government's tightening control over developers' borrowing, real estate developers had to resort to non-standard ways of borrowing. They used various financial conduits to raise their own funds, such as corporate bonds, senior notes, joint values and strategic alliances, perpetual bonds, trust firms, and, more recently, real estate investment trusts (REITs). However, the last of these is a private equity fund rather than a standard public one.

In addition, there are so-called joint ventures and strategic alliances that might not be considered equity finance. This can instead be debt financing, known as 'borrowing capital in the name of equity.' The developer aimed to gain capital and signed an agreement with the lender for a fixed income, which is, in fact, a debt. In this case the lender may not be a financial institution that offers loans based on land property guarantees. However, the partnership gives the partner access to the landed property. Because it is disguised as equity finance, the capital is not treated as a land mortgage. But it is de facto land-based financing.

De-financialization

The second stage of financialization is 'de-financialization', which expands more regulated financial products, such as the corporation bonds of UDICs (*chengtou*) and local government bonds (LGBs). Here, 'de-financialization' does not refer to the removal of financial products. Rather, it means a more regulated process through financialized statecraft. De-financialization means that financial logic is subjected to state control to contain financial risks (Li et al. 2023b). This means further treating these financial products according to the market operation rules by removing local governments' influence, i.e., implicit guarantees for repayment.

The purpose was thus shifted from capital mobilization to risk containment. However, in order to contain the financial risk, the state needs to raise capital—but not purely for development. Hence, this is a critical difference between the developmental state and financialized state entrepreneurialism because the latter, at this specific stage, is statecraft with financial tools. The statecraft is illustrated by deleveraging the real estate developers' debt and creating a formal, regulated financial product—local government bonds.

De-financialization involves two major measures. First, deleveraging the debt of developers. By the mid-2010s, the real estate boom had generated many unsold commodity housing units, reaching a nationwide total of 739 million square meters. The situation was particularly serious in small and medium-sized cities.[3] With the unsold housing stock, real estate development companies had a high debt ratio, which might further implicate banks. To cope with the financial risk, the central government initiated a policy to 'deleverage' developers' debts in 2015. The policy boosted housing demand in small and medium-sized cities through the 'shantytown renewal program' (He et al. 2020; Wu et al. 2020; Wu 2023a).

This uses the National Development Bank, a policy bank, to provide initial capital. The development projects compensated residents in the designated demolition area, encouraging them to buy new housing in the renewal project or elsewhere. This effectively increased the demand for housing and reduced the unsold housing stock. The latest property boom lured more residents into investing their own financial resources in housing (Wu et al. 2020). The policy reduced the down-payment requirement and housing transaction taxes to encourage housing purchases. As the mortgage-to-household income ratio was still relatively low in China, the deleveraging policy stimulated the use of housing mortgages to speculate the housing boom, which essentially reduced the financial risk of the real state sector through household debts. The large-scale shantytown renewal program started another wave of financialization —the financialization of urban redevelopment (Rutland 2010).

However, the process of financialization was not initiated by real estate developers or the financial sector to capture financial benefits by turning public housing into financial assets. The state started it to cope with earlier credit expansion and its consequential financial risks. The state was directly involved in this process using the policy bank, financial instruments, and urban renewal programs.[4] At the time, household debt, especially in housing mortgages, was still considered safer than development companies and local government debt, which was implicated through hidden financial guarantees. The property boom driven by financialization benefited developers and speculators, but aggravated the housing affordability problem. Developers aggressively borrowed capital, obtained land, and used the pre-sale or pre-completion deposits for homes bought 'off plan' to make a quick turnover in the housing market to make a profit. The deleverage policy rescued high-debt developers, but did not solve the systematic financial risk. It deferred the risk into the future and even laid down further risks, such as the bankruptcy of Country Garden using this business model.

Second, regulating the local government debt. Through LGFVs, the local government debts escalated. The exact amount of debt was unclear, as there were hidden financial obligations. The debt was estimated to reach 50 trillion Yuan (Wu 2023a). Since 2014, faced with the mounting local government debt and financial risks brought about by implicit guarantees and off-balance sheet borrowing through shadow banking, the central government has initiated stricter financial regulation. To regulate the local government debt and ease the financial pressure, the National People's Congress promulgated the revised Budget Law, allowing the local governments to use 'local government bonds' to finance infrastructure construction (Li et al. 2023a). The bond provides a new source for local governments to borrow formally from the financial market. It is more formal and regulated because the State Council regulates bond issuance through a quota system, and the provincial government is responsible for issuance and repayment.

The debt–bond swap program clarified the financial obligations and debts under implied guarantees provided by the local government and converted them into bonds with lower interest rates. In 2015, the new regulation prohibited using bank loans or other financial instruments for land purchase to speculate land appreciation. The status of UDICs is clarified as an enterprise without government financial guarantees. Land reserve centers are transformed into public land authorities and can only use budgetary revenue to finance land acquisition. The policy prohibits land reserve centers from acting as LGFVs (Wu 2023a; Feng et al. 2023).

Through these formalization policies, the hidden debts are made explicit. It is estimated that the formal or 'explicit' local government debt in 2019 was 27 trillion Yuan, accounting for about 30 percent of GDP (Wu 2023a). The central government insists that through this formal procedure of classifying local government debts, the central government is no longer responsible for local government debts, hoping to control the escalating local government debt.

As mentioned earlier, the policy of deleveraging the real estate sector's debt resorted to a financial approach, i.e., stimulating housing demand through monetary compensation for demolition. This triggered a property boom, especially in low-tier cities. The policy solved the unsold housing stock of some developers but, at the same time, stimulated further borrowing and increased debts.

The central government issued a 'three red lines' policy in 2020 in an attempt to prevent further financial crises. It sets the limit to which developers could borrow further: (1) their debt-to-asset ratio is below 70 percent; (2) their debt should not exceed equity; (3) the cash covers the entire short-term liability.

Developers who meet all these restrictions can only increase their debt by an annual upper limit of 15 percent. For meeting two restrictions, an annual increase in debt cannot exceed 10 percent, and for one restriction, no more than a 5 percent increase in debt. The control over financing led excessively indebted developers, such as Evergrande, into a difficult situation. The policy and COVID lockdown reduced both housing demand and the level of confidence in the property market. The liquidity problem has worsened since 2023.

In sum, the history of financialization demonstrates that the Chinese state promotes a financial approach to urban development and focuses on crisis management. The process led to a significant financial expansion, diverse financial conduits, and entangling state–finance relations. The local state absorbs financial resources through its LGFVs, and developers adopted a debt-driven approach to capitalize on the property boom in the aftermath of state-led financial expansion. However, the constant financial risks also promoted the state to 'de-financialize,' imposing a more regulated approach and separate state-finance entangling (e.g., issuing local government bonds and separating the bonds of UDICs and government debts). The following sections will further explain these financialization and de-financialization processes in housing, land, and public finance.

Housing commodification and assetization

The starting point of financialization is housing commodification, which makes land valuable. Land development further allows local governments to extract value to finance infrastructure development.

In China, commodification is widely used in housing policy rather than the concept of financialization. Public housing provisions create a heavy financial burden for the state. China has promoted homeownership through commodification; all new housing takes the form of a commodity through private consumption. The government managed to get rid of funding pressure. For households, housing has become an important financial asset. The ownership rate in China is significantly high—over 80 percent of urban households are homeowners. In this sense, China is a nation of homeowners, at least in its cities. Housing purchase absorbs Chinese households' savings. Despite the huge financial pressure on households, they are willing to bear the burden because they expect value appreciation. There is also a strong demand for housing during fast urbanization. Households strive to improve their living conditions by purchasing new houses. Household savings fund housing consumption. To a lesser extent, it is financed by bank loans or securitized forms.

In addition to the need for larger and better space, households treat housing purchases as an investment. Housing as a financial asset provides a capacity for future needs, such as children's education, healthcare, and additional pension income. In the literature, households' investment in individual welfare has been called 'private Keynesianism' (Crouch 2009; Forrest and Hirayama 2015) because households look after themselves through their own financial capacities rather than state funding. Housing is no longer a financial burden for the state. China has carried out large-scale housing construction since the introduction of housing reforms. Because households expect an appreciation in value, they are willing to invest in housing construction.

China's housing development is also associated with financial changes. Its housing policy changed because of a series of financial crises. For example, in 1997, the Asian Financial Crisis forced China to abandon its public housing system, beginning the large-scale commodification of housing. Since then, there has been a long property boom (Wu 2015a), although the housing market has fluctuated from time to time. There has been a strong expectation that house prices would continue to rise. This is because China is experiencing a period of rapid urbanization. The supply of land is limited through the operation of state control. The expectation of future price increases motivated households to invest in housing as an asset.

Although housing prices have fluctuated, the overall trend has been price inflation. Since the Asian Financial Crisis in 1997, China's housing market has entered a boom. The government imposed various restrictions on housing sales to control the speed of price increases. The restriction temporarily slowed down the pace of housing price increases. However, housing prices have still experienced significant growth. The market reached a peak just before the global financial crisis in 2008. After this, the house price dropped quickly but picked up swiftly under the stimulus package, which led to capital flow into the property market.

In market economies, housing financial research focuses on housing mortgages. Housing mortgages in China are an important source of finance, but they do not occupy a central position in China's housing financialization. In the early years of housing reform, the level of mortgages was relatively low, and households were reluctant to borrow (Li 2010). Increasingly, however, housing mortgages have been used and have become an important source of finance for homeownership. The level of household debt also increased. Households' savings rather than mortgages have driven the commodification of housing. In the early stages of housing reform, households used their savings to finance their homeownership.

Later, when mortgages are used, the mortgage is financed by banks through loans, mainly from household deposits (savings) instead of the capital market. The level of securitization was relatively low. The banks did not sell their mortgages to other financial institutions. So-called mortgage-based securitization (MBS) or asset-backed securitization (ABS) were relatively new and underdeveloped vehicles. Instead, banks used household savings to lend mortgages to households. The housing provident fund has been set up to promote the housing market. It assisted housing purchases for those in the public or the formal private sector. To constrain the level of real estate speculation, the Chinese government imposed various restrictions on the use of commercial mortgages. Accordingly, mortgages have not run out of control. Households use their financial resources to finance housing consumption. With increased house prices, households now have to use commercial mortgages, perhaps in addition to housing provident funds, in order to finance homeownership.

Mainland China has a unique house pre-sale system for financing housing development projects, similar to Hong Kong. The pre-sale practices allow developers to get funds before they complete housing construction, which considerably accelerated up the circulation of financial sources. The pre-sale can constitute about 40 percent of the total finance. Households will often put down a deposit on a new property before it is completed. In residential property development, the developer usually adopts fast turnover tactics. Thus, it only needs six months of bank loans. Within this short period, the developer tries to complete the foundational work and then can sell the housing property to buyers.

In addition to development loans, the developers must raise capital on their own account. A large proportion of housing finance comes from developers' 'self-raised capital' because the state has tightened the regulation for real estate investment. The developer strived to expand the financial source. One way is to use 'debt in the name of share,' in which the developer tried to find a business partner to develop the project as a joint venture. The partner, as an equity investment, provides capital. However, the partner is an investor because it has a contract to invest in the debt disguised as equity.

The question is why the developer is interested in using this approach. Investors escaped from financial regulation—they lend more than their balance sheets because their lending is not treated as a debt. The form of debt in the name of shares is forbidden because the financial organization lends out more money, disregarding the financial risks because they treat it as an actual debt to be paid back by the borrower. In short, for the developer, there is a relatively tightly regulated environment. It tried to use the pre-sale system and the debt in the name of shares in order to increase its leverage for finance. This is profitable in a low-interest rates environment and during periods of asset inflation. The tactic is a quick turnover using other financial resources.

In short, housing financialization is, therefore, a process of assetization.

The meaning of housing financialization in Western markets often refers to housing mortgage securitization and its techniques, such as credit ratings and subprime mortgages. Social rental housing even becomes financialized, for example, through real estate investment trusts (REITs) (Risager 2021; Aalbers et al. 2023). In contrast to Western economies, where housing mortgage securitization by the finance sector is a widespread practice, China does not rely heavily on securitization. China's housing financialization is more related to changing housing policies, such as housing commodification, which encourages households' investment in housing.

We must also consider the changing approach to financing housing development projects, including housing pre-sales and shadow banking practices.[5] Financial organizations lent money to the developers outside the balance sheet under irregular (non-standard) financial products.

Housing financialization, in China's case, means that housing is turned into valuable assets, not for institutional investors. It is mainly for ordinary urban households. Their investment in housing is important for financing urban development in China. That is, financialization in China is a process of assetization. For households, housing becomes a valuable financial asset; subsequently, land becomes an important financial asset for both the state and developers. Indeed, since 2008, land-based urbanization has become an essential driving force for China's economic development.

China also echoes the notion of 'rentier capitalism' (Penny 2022). The financialization of rental housing means that external investors seek housing as an investment item to gain profits from rentals. In New York City and Berlin, weakened rental protection helps to transform affordable housing into a 'new global asset class' (Fields and Uffer 2016). In the UK, housing companies are established by local governments and involved in 'local state rentiers' (Penny 2022). In China, housing became an integral part of capital accumulation. this represents a state-engineered process. Instead of using external investors, the policy encourages households to invest in housing through their savings. Therefore, housing financialization is linked to the changing housing institutions and statecraft that make housing both a consumption and investment product. It mixes 'managerialism, entrepreneurialism, and financialism' (Pike et al., 2019, p. 29).

Financialization as a statecraft means that the application of financial means is conditional. Facing property speculation, the Chinese government announced a tightening of the rules around housing purchases and mortgages, as 'housing is for living, not for speculation' (Wu et al. 2020). Recently, the government has introduced a shared-ownership housing scheme. In contrast to extending financialization in social rental housing of low-income groups in Western economies, the scheme excludes low-income households for fear of financial risk (Shen et al. 2022). Hence, the policy limits rather than extends financialization, which can be considered part of the process of overall de-financialization in China.

Land finance and land financialization

Land has been a key instrument for entrepreneurial local governments to attract investment. Through the compulsory acquisition of rural land at a lower price and the sale of land development rights at a higher price in a competitive market, the local government managed to make a profit. This is known as 'land finance' or 'land-based finance,' which explains that the revenue derived from land is an important source of public finance (Tao et al. 2010). This fiscal perspective suggests the importance of the real estate sector to the state's fiscal condition, i.e., the state making money from land (Tao et al. 2010, Lin 2014). The other aspect, however, regards how the local state uses land to raise capital for urban development: that is, land financialization to finance urban and economic development. Based on China's unique land institution, land financialization is a specific land-based statecraft to promote economic growth (Lin and Yi 2011; Wu 2022c, 2023a). Therefore, the objective of land finance goes beyond revenue generation and helps to achieve more strategic goals of the state, such as macro-economic development.

From the financialization perspective, we can understand how the land has been financialized. The production of land is motivated by profit-making, but the shift toward large-scale land financialization is associated with China's changing development in the late world-factory model. The global financial crisis in 2008 marked a new wave of land financialization beyond revenue generation because land development became less profitable with an increasing cost of land acquisition (Wu 2022b). It was observed that local officials were keen to use the financial platform built upon land assets in order to raise capital rather than care for land prices, and the development has shifted from possibly profitable projects to projects with great uncertainty.[6] These investment platforms extended from the provincial levels to township levels. The system of investment platforms became chaotic. The local governments find a way to register capital on these platforms and then use the land to raise capital.

Hence, we witness a shift from land finance—making a profit from land—to land financialization as changing statecraft. There is a large body of literature on land financialization in the West. Perhaps David Harvey (1982) was the first to treat land from a financialization perspective systematically, when he proposed the theory of the 'urbanization of capital.' Capital is invested not only in the production sphere but also the built environment, such as through the process of 'urbanization.' Investing in land is the 'urbanization of capital.' Because of this combination of land and capital, or capital taking the form of land, land is treated as a financial asset.

The financial characteristics of land have been exposed since the subprime mortgage crisis (Gotham, 2009), when these were traded as assets on the financial market. Thus, the house is a living place and a financial asset traded in the financial market. This financial trait is also revealed in the system of 'rentier capitalism' because landlords rent the properties for rental gain (Christophers 2010). Similarly, in East Asia, 'land monetization' is an important feature (Shatkin 2017). Even the state is involved in this process of land monetization. Singapore has become a 'property state' (Haila 2016) because it is interested in the property market and regulates it through urban policies. In Western market economies, land financialization is the result of financial deregulation and the introduction of new financial techniques invented to facilitate capital mobility. This created the 'great wall of money' (Aalbers 2008), circulating the globe and seeking investment opportunities. To investors, housing mortgages are high-quality collaterals. It is a secured form of investment to generate steady returns. Real estate investment trusts (REITs) are now interested in rental markets. Social rental housing was traditionally outside private financial interests, but it has now been financialized. Under the new deregulated financial market, investing in social rental housing became attractive (Fields and Uffer 2016; Van Loon et al. 2019).

For China, borrowing from land means that the land mortgage is now an important source of development finance. There is no nationwide data with regard to China. However, using a land mortgage is a widespread practice. The Ministry of Land and Resources organized a survey of 84 cities in China and found that local governments used 1.8 trillion Yuan of land mortgages. By 2014, this figure had been inflated to 9.5 trillion Yuan (Wu 2022c). This suggests that land mortgages have increased significantly after the global financial crisis, which is related to the requirement of the fiscal stimulus package for local matching funds (Liu et al. 2022; Li et al. 2023a). The local governments set up the land reserve centers to initiate the system of land banking. The function of these centers was to use land to borrow loans from banks. These practices became widespread as a new landscape of financialized development.

Because of the concern about financial risks in this area, the land mortgage was suspended in 2014. The problem is that the land reserve centers used the 'reserved land' to borrow capital. Local governments used the land to gain the mortgage before the land received the actual title (of saleable land). According to the Chinese land system, the land has to go through a public auction to convert it from rural land to state-owned land, through which a land deed is created. The land market reveals the value of land through land transactions. Otherwise, the value of land is difficult to judge.

The Chinese case of land financialization reveals the contrast between a fiscal perspective of land finance and a financial perspective of land financialization. The former indicates profit-making from land due to the growth machine, while the latter highlights the statecraft of using land to mobilize capital. Land financialization means less land value capture and the generation of income from land value appreciation. This is more about speculative state actions to convert the land into 'collaterals' for capital mobilization.

In this way, the land is treated as 'fictitious' capital because its value is not realized or probably not realizable because of this speculative nature. Using the land before it gains the property title is a practice of 'virtual capital circulation' (see the earlier discussion on the history of financialization in this chapter). The state started without cash, but through land, it gained the credit for capital to carry out development. However, this process does not always lead to value appreciation. There is no guarantee that this could be fulfilled. The local government may waste investment in land development, which does not contribute to economic growth or social benefits. This eventually led to greater financial risks.

To a larger extent, China's land value appreciation is attributed to an overall increase in monetary supply and a tightened land supply control. If land development is combined with the real economy, such as export-oriented industries or domestic consumption, this contributes to GDP growth and tax income. However, in the aftermath of the fiscal stimulus package and more accessible liquidity, capital mobilized from land tended to be spent unproductively and speculatively. The land mortgage thus became the local government debt, which is not payable within land projects.

This can be seen from China's land mortgage survey. The survey indicates that 14 percent of land mortgages were based on land reserves that could not be sold. The survey also shows that the land mortgage ratio has increased because of land value appreciation, which means more capital can be borrowed from the same amount of land. In this system the amount of borrowing is based on expected land value rather than the actual land price. Thus, there is an increase in the intensity of borrowing from the land. Over 75 percent of land mortgages were provided by major state banks. In other words, the state banks provided capital to local governments based on expected land value. This is a more speculative behavior—hoping a land value increase would cover the debt. Land mortgages are thus different from usual development finance, distinguishing land financialization from land finance. The latter provides capital for a development project. However, land financialization is more speculative, and the hope is to capture future land value through debt finance, which is riskier.

As seen in the land mortgage operation, China's land financialization can be characterized as 'state-led financialization.' This means that the state enables other financial actors to treat land as fictitious capital, but is also directly involved in creating it by speculating state land assets. First, the state captures land by declaring that it is a state asset—even for rural land, its development right is subject to strict constraints of the nationalized development right. Second, the state creates a competitive land market by financializing land development rights. This includes an assetization process through which the development and investment platforms use the land to raise capital. In this way, China's land financialization is state-centered.

Our explanation of land financialization places housing, land, and infrastructure into a single framework of state-led financialization. These aspects are interlocked and pave the way for each other. Housing financialization mobilizes households' savings into treating the property as an asset for them, contributing to housing construction and, in terms, creating a land value. Land financialization is built upon the statecraft and land institutions to anchor land for capital liquidity, through which the state manages to provide credit for urban infrastructure development. Expecting future land rent justifies the land value, which allows local government financial vehicles or development corporations (*chengtou*) to borrow (Feng et al. 2022a).

Public finance and infrastructure development

The final aspect is infrastructure development and public finance. Infrastructure development brings together housing, land, and urban development. The key actors are urban development and investment corporations (UDICs), or *chengtou*. These are state-owned enterprises that specialize in land and infrastructure development. They are affiliated with various levels of government and belong to the State-owned Assets Supervision and Administration Commission (SASAC) of the State Council. As state-owned enterprises (SOEs), they are given exclusive rights in the primary land development, which converts 'untitled' rural land into saleable urban land. A UDIC is designated as the developer for a specific area, while other UDICs and private developers can participate in the secondary land development. This designation by the local state actually endorses the status of UDICs and provides credit to them as an 'implicit' government guarantee. In this primary land development, the UDIC is a franchised operator. It then borrows capital from banks. Overall, UDICs' business is strongly related to real estate development, and the real estate part makes their business profitable. Although they may not conduct property development themselves, they are able to sell the land to property developers at a much higher price. This means extracting value from land. It is a business model of UDICs.

However, as SOEs, they may not be solely profit-oriented as they receive various state interventions, not through market regulation but more directly through their state ownership (Wang 2015; Feng et al. 2023). Sometimes, they should carry out unprofitable infrastructure development for the local government. Indeed, there are various types of UDICs. Some are more functional, for example, to achieve the assigned goals by the state to improve infrastructure quality or upgrade economic structure. Some are more business-oriented, probably operating in the property market themselves.

In short, UDICs put land, housing, and infrastructure development together. They are not simply LGFVs, as they undertake the development functions. But they are government development arms and financial actors for the state. Through their operation, they are state agents and incur the debt. Because of this association, their debt has been regarded as the main part of local government debt. The new financial regulation, introduced in 2015, hoped to separate them from the local government. By reimbursing the previous debt incurred by UDICs for the local government, the new policy forbids local governments from raising capital through UDICs. They are now treated as an enterprise. Their debt is corporate debt. UDICs are allowed to issue bonds, which are not treated as local government debts.

UDICs are development actors but also serve as LGFVs. They carry out the primary land development for the local government. However, their status has changed since the onset of the global financial crisis in 2008. They have been turned into LGFVs to borrow money. Some LGFVs may only take financial functions and may not undertake actual development. Thus, LGFVs include state-owned asset management corporations and industry-specific investment corporations—for example, a government-guided industrial investment fund (Pan et al. 2021). The example mentioned in Chapter 1 about Lingang Economic Group belongs to the industry-specific development corporation but also takes into account the function of investment and land development. Local governments used to guarantee their loans, which is now forbidden.

Injecting land into the development corporation effectively guarantees a 'franchised' status for a designed area. Using bank loans to fund initiate land preparation is problematic because the loan is a short-term one, while infrastructure development is carried out over a long term. The real problem is that it was unclear whether this was a corporate or a government loan. This issue has now been clarified. The debt of LGFVs was regarded as a local government debt. Officially, LGFV's debt amounted to 22.9 trillion Yuan in 2014 as official lending, but the liability could reach 47.7 trillion Yuan (Bai et al. 2016), raising a significant concern for local government debt. However, after the new financial regulation, the government hopes to eliminate this debt.

By investigating UDICs' behavior in the land market, we can see that they do not directly generate revenue or profit. It is less about contributing revenue to local public finance. They are important because they expand the capacity of local states for capital mobilization. They only paid the local governments between 4 and 5 percent of land conveyance fees (Wu 2022b). They do not contribute significantly to local government income. But they can access the capital market for funds. They construct infrastructure that sets the conditions for the property market, contributing to the local government's land income, and property developers also contribute to tax income.

To sum up, China has seen limited housing mortgage securitization. This differs from financialization, as seen in the subprime mortgage crisis. However, China has seen its financialization through real estate development. Financing urban development in China increasingly resorts to the financialized approach. This financialized way of urban development has led to the financialization of the entire Chinese economy (Wu et al. 2022a; Zhang and Wu 2022a; Theurillat and Graezer Bideau 2022). This chapter investigates several aspects of financialization in housing, land, and infrastructure development.

Built on the Chinese households' aspiration for better housing, the state's monopoly of land supply, and an illiquid financial environment controlled by the state and state-owned banks, housing commodification in China has mobilized households to invest in housing as financial assets for value appreciation (Wu et al. 2020, p. 1494). During a housing boom, the return far exceeds the interest from bank deposits. The households are willing to take a debt to finance homeownership. Although Chinese families tend to have a higher saving rate, household debt is rising rapidly. The source of housing finance came from bank deposits rather than debt securitization.

The Chinese pathway converts bank savings into housing assets, known as housing assetization. This means China's monetary supply is homegrown and does not originate in the global financial markets. Housing financialization is not the penetration of financial capital into housing through debt securitization (Aalbers 2016; Fernandez and Aalbers 2016). The latter tends to privilege financial actors and practices. In contrast, China's home-grown housing assetization is characterized by the state's capacity to provide monetary supply based on housing and land.

The key agents of financialization in the Chinese case are the state and the millions of homeowners and mortgage borrowers. Thus, Chinese housing financialization is focused less on financial sectors. Rather, the features of China's housing financialization are a high level of owner-occupation and also the willingness of Chinese households to endure high levels of indebtedness and financial burden to gain future asset appreciation. This is similar to a model of asset-based welfare in the West (Rossi 2017).

China's housing financialization is based on rapid economic growth in the post-reform period. Housing financialization created an outlet for surplus capital, which was not fruitful in the manufacturing industries, where domestic demand lagged. China's financialization is not a finance-initiated process. Rather, it relies on creating a new commodity housing sector, which plays a vital role in credit creation.

Instead of thinking of China's housing financialization as a process subordinate to absorbing global mobile capital, it is driven by the need to create a new investment sphere in which new capital is created. It reflects the governance approach of 'state entrepreneurialism' described in this book. Because of housing assetization, a new land-finance model has been created, which further facilitates a broader shift in the Chinese economy toward making money and capturing value not through producing goods and delivering nonfinancial services but through the holding of property assets (Wu et al. 2020, p. 1495). In this regard, it conforms to the definition of financialization (Krippner 2005).

Nevertheless, the dynamic is more 'indigenous,' being generated from its contradiction of capital accumulation, through which housing financialization helped to break out the more restrictive and depressed financial environment. It created a de facto interest rate liberalization, as households are also customers of wealth management products that build upon property assets. The housing pre-sale system also provided a crucial financial source for property developers. The state's restriction on housing sales and control over their financial process becomes less effective, as developers can resort to various practices of shadow banking to raise money outside the bank's balance sheet. This Chinese assetization is similar to the system of land banking and asset acquisition in the West (Ward and Swyngedouw 2018). The latter means large firms can raise finance through assetization, co-existing with the penetration of financial capital (Fernandez and Aalbers 2016).

Because China's housing and land assetization originated from millions of households and state-owned development corporations, the implication for governance is that bondholders' value does not impose on the process of financialization. In the West, through imposing a financial logic, financialization is an 'important material mechanism driving neoliberalization' (Ward et al. 2019, p. 124) because asset owners mobilize capital to finance their projects. In China, it is a state-engineered process—creating housing assets allows the state to inject liquidity into urban development. As such, housing is more than the bearer of financialization (Aalbers 2016). It is an instrument for credit creation and, in turn, a financialized economy.

Housing financialization materializes the bid land price. The value is captured and transferred into housing prices. It allows more capital to be mobilized. Thus, housing financialization plays a key role in the overall financialization of the Chinese economy. Housing financialization is a key process through which the state can mobilize capital. The financialization process is a state strategy that motivates millions of enthusiastic homebuyers to fix the over-accumulated economy. However, such a strategy is speculative because it creates a high debt-to-asset ratio. The asset has limited or virtually low rental income, which does not justify its value as a financial asset. It is based only on continuous value appreciation or on asset inflation. Once the inflation ceases, there will be a looming financial risk.

What exactly is meant by 'financializing Chinese cities'? Implications for Chinese economy and governance

From the above discussion, we can see that China finances urban development through the adoption of a financial approach. Does this application of financial methods mean that Chinese cities are financialized? In contrast to rising finance facilitated by the state, the Chinese process is more state-centered. China's urban development was financed mainly through bank loans by development agencies such as urban development corporations. Similar to ordinary development finance, this does not precisely mean 'financialization,' although banks have become more important in urban development than they were during the planned economy era (Jiang and Waley 2022; Shi 2023; Wu, 2023a). The land development has led to land finance, creating a source of revenue for public finance.

The process of financialization began during the global financial crisis in 2008 when China initiated a 4 trillion Yuan stimulus package. Since then, in order to cope with the pressure of loans, China started a more financialized process, for example, using non-standard financial products—shadow banking—outside the balance sheets. Urban development is increasingly financed through financialized means. Land mortgages converted land into financial assets, and land further formed the basis for other financial products. Although land mortgages on untitled land (land reserve) were banned in 2014, the practice has already created a link between the debt of development corporations and local government debt under various implicit government guarantees. The state attempted to control the local government debt by regulating and separating UDIC debts from local government debt. Nevertheless, controlling the financial practices in urban development finance does not reduce the scope of financialization. Instead, the regulation aims to prevent shadow banking by introducing two more regulated financial products—UDIC (*chengtou*) bonds and local government bonds (LGBs) (Figure 4.2).

In this sense, Chinese cities are indeed financialized. UDIC bonds have grown significantly since the 2010s. The bond constituted 75 percent of total enterprise bonds in 2017; in 2014, the outstanding bond reached 4.9 trillion Yuan (Wu 2022b). As *chengtou* bonds are mainly used in urban development, from the figure, it can be seen that many financial sources had been diverted into the built environment, a process described by the shift of capital circulation (Harvey 1978). It also suggests that urban development has become a major driving force for the Chinese economy. Land, as the asset of development corporations, is used for bond raising. Issuing *chengtou* bonds relies on confidence in the real estate market.

Land as financial assets and *chengtou* bonds based on land created a new condition for financialization. In 2016, over 62 percent of *chengtou* bonds were invested in wealth

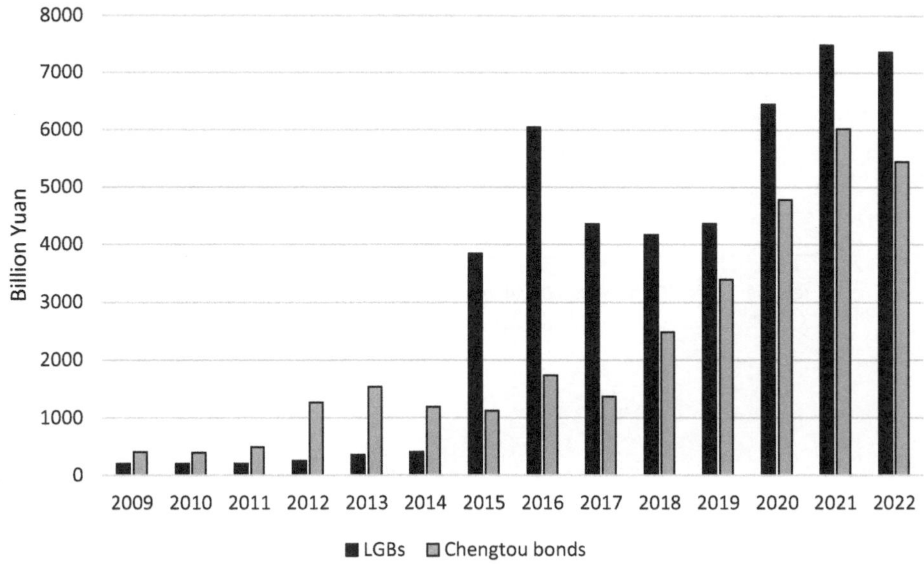

Figure 4.2 The increasing *chengtou* bonds and local government bonds in China.

Source: Li et al. 2023b.

management products (WMPs). Inexperienced ordinary households buy these WMPs. This conduit changed the conventional route for households' saving deposits in banks, which in turn provided loans to development corporations. Thus, the financing of urban development in China takes more financial means.

Figure 4.3 shows a complex picture of financial conduits in the aftermath of financialization in Chinese cities; in addition to the conventional direct finance, which is a characteristic of the Chinese financial system, new securitized and financialized conduits are created, including housing and land mortgages, local government bonds (municipal bonds), *chengtou* bonds and WMPs and various other non-standard financial products and financial deals under irregular practices (such as 'fake equity, real debt,' as mentioned earlier in this chapter).

The changing urban development finance has a significant implication for the Chinese economy. The financing urban development through fictious capital has led to an overall trend of economic financialization . The financial conduits created along with *chengtou* and land development allow Chinese cities to access financial markets. The economy is 'urbanized' through the 'urbanization of capital,' namely the diversion of investment from manufacturing industries to the built environment. The landed properties (housing, land, and infrastructure) experienced investment-driven asset appreciation. Assets become more valuable, which further justifies credit expansion through 'borrowing to finance development' (Pan et al. 2017).

The new financial approach allows the state to add other objectives to finance non-profitable programs. The state uses a financial rather than a fiscal (welfare) approach to stimulate the economy and manage crises. Land reserve centers created land as financial assets, initially with bank investment and now through local government bonds. Based on land assets, LGFVs

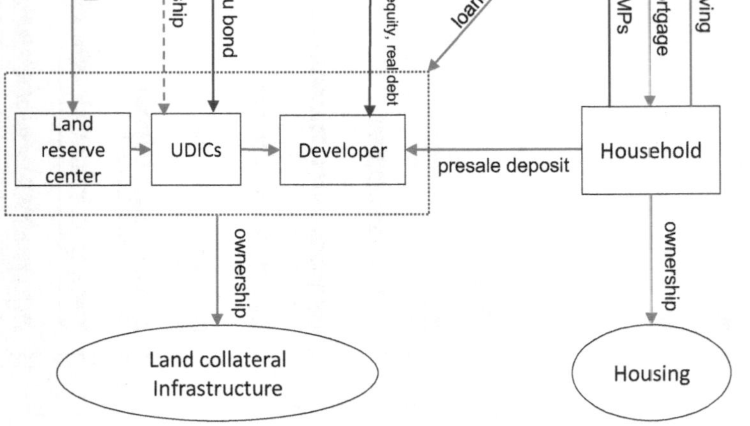

Figure 4.3 The complex financial channels in China's urban financialization.

Source: Wu 2023a.

raise development loans and *chengtou* bonds. Together, they created an internal money creation process, as shown in the rapid expansion of monetary supply (M2). Investing in 'fixed assets' is a major economic driving force. In contrast to a relatively lower level of consumption, exports and investment have been two pillars of Chinese economic growth.

This specific form of financialization has implications for governance. As it is a process of state money creation, there is a tendency to increasingly use fictitious capital in urban development at the expense of the real economy, as reflected in the Chinese expression of 'from reality to virtuality' (*tuo shi xiang xu*). The centrality of governance does not shift to the finance sector. This contrasts with the increasing influence of bondholders' value in late capitalism (Peck and Whiteside 2016).

More recently, there has been an effort to regulate financial practices and constrain the financial logic. The central government issues the quota of local government bonds, and the provincial government is responsible for issuing and paying back LGBs, which are tied up with either an income stream from specific projects for special bonds or fiscal income for a general bond. The attempt is to anchor the bond with a real income. Therefore, although bonds are the means for financing urban development, their issuance and operation do not reflect an overwhelming finance sector. The state is central to the process of financialization. Financing urban development through financial means does not automatically mean financialized urban governance based on the financial sector. It reflects the statecraft of 'state entrepreneurialism,' as we discuss in Chapter 2.

Critical reflection

The Chinese case shows that financialization is a state strategy to cope with the impact of external financial crises (Wu 2023a). It is a by-product of state operation. This is not dissimilar to other situations. For example, the Canadian state treats surplus public land as a financial asset and operates through financial approaches (Whiteside 2019). Similarly, in the UK, the local state initiated financialization as a statecraft in response to austerity (Pike et al. 2019; Christophers 2019). In China, raising capital for state development projects is important. The credit expansion, as shown in a significant increase in monetary supply, is anchored in landed properties. As such, there has been a shift from manufacturing development driven by foreign investment to urban land development driven by indigenously created investment. The credit expansion after 2008 was feasible because of the financialization of land. It creates a new space for capital accumulation. The monetary supply and investment in fixed assets are coupled, showing significant expansion until the recent pandemic. This reveals the capital flow from production to the built environment.

While Harvey's theory of capital circuits already predicts the switch between primary and secondary capital circuits, the mechanism is less known (Harvey 1978). The theory of capital switch has also been discussed in China (Wu 1997) without the concrete pathways being known. His theory assumes a declining rate of profitability due to over-accumulation in the production sector. The role of the state in his theory is less prominent. He uses the term 'spatial fix,' suggesting that this could be a possible state action as part of its strategy for class restoration. The most salient operation of this class restoration is 'accumulation by dispossession,' often associated with a new financialized development (Harvey 2003). In the literature on financialization, the role of the state is frequently mentioned as 'enabling'— through deregulation or other neoliberal means (Gotham 2009; Pike et al. 2019; Weber 2010; Whiteside 2019). Some actions are more proactive than reactionary, characterized as 'statecraft' (Lauermann 2018; Pike et al. 2019). This chapter explains how the Chinese state manages to overcome various institutional and financial barriers to achieve this spatial fix and a coupled outcome of this process as capital switching.

First, until its revision in 2015, the Chinese Budgetary Law did not allow local governments to borrow from the capital market. Second, China's financial system is characterized by a bank-oriented system. It was subject to a financial depression rather than deregulation or liberalization. Simple quantitative easing is difficult to achieve. These obstacles were solved by using local government financial vehicles. As state-owned corporations, they borrowed on behalf of the local government to invest in major infrastructure development and, to some extent, affordable housing.[7]

This initiated a process of financialization as an evident financial expansion. Through financial expansion, profits are made from financial means and land development. This meets the first criterion of financialization, as profit-making is not limited to production and trade but through financial conduits (Krippner 2005). For land development, profits are made not only through the production of landed properties but also through land value appreciation and assetization—turning land into financial assets through collateralization. Land development proceeded even with a small profit margin because the main purpose was to raise capital.

In addition to the state's role, it is important to reveal the complex motivations for using financialization as an entrepreneurial statecraft. They include financing development, refinancing debt, and deleveraging financial risks. This means that financialization is a governance change. It occurs because the state changes its development operations, for example, using development corporations and separating corporate bonds and government bonds to contain financial risks and pursuing objectives other than a narrowly defined financialization purpose, for example, using green bonds for the low-carbon economy.

As distinguished in this chapter, financialization contains financial expansion and strengthened financial logic. China's financialization often confirms that it is a financial expansion but that it does not adhere to financial logic. To comprehend the logic of financialization, we need to consider the state and the crisis arising from capital accumulation. China shows that the process of financialization is triggered by capital accumulation, but that its materialization is beyond the capital accumulation itself. It is a 'spatiotemporal fix'—centered upon the state. Nevertheless, the result is that the state is even more deeply constrained by capital accumulation rather than free from the financial logic because there is an ever-increasing risk and crisis. The state deploys financial means while half-heartedly embracing the financial logic, which is the root cause of the financial risk in China.

This chapter shows that the state is important in enabling and constraining financialization. The latter may be referred to as 'de-financialization.' However, such a de-financialization process is accompanied by financial expansion, as shown in the expansion of *chengtou* and local government bonds. While the role of the state in financialization has been widely noted in different contexts in late capitalism and emerging markets (Halbert and Rouanet 2014; Klink et al. 2020; Goldman 2020), China's financialization originates from the state's deployment of financial approaches to urban development (Wu et al. 2022a). It is a process of internalizing financial logic in the state development system without fully adopting the financial logic, a feature of state entrepreneurialism. The expansion of financial operations partially meets the criteria of financialization. However, the private financial sector does not initiate financialization as a unidirectional process to subject urban development under a financial logic. Instead, it reflects the state's developmental intention and extensive state involvement and participation in finance. It simultaneously evolves into greater interference and constraining the financial logic—known as 'de-financialization' (Wu et al. 2022a; Li et al. 2023a).

There is a long tradition of research on financing urban development. The 'urbanization of capital' theory can be understood as land financialization because land is treated as a financial asset (Harvey 1982). The theory describes the arrival of capital in the city. The capital penetrates the city and subjects it to the logic of financialization. The process is also a matter of negotiation and interaction between the finance and the state (Theurillate et al. 2016; Aalbers 2023). The question raised in this chapter is, by whom is the city treated as a financial asset? Housing financialization literature suggests the Great Wall of Money was created by financial deregulation and securitization of mortgages (Aalbers 2016; Gotham 2009).

Similarly, this chapter shows that the political economy of post-reform China confirms that land has been assetized. Land value capture is used to finance urbanization (Figure 4.4) (Feng et al. 2022b; Wu 2022c). However, treating land as a financial asset is a state strategy. The land is made a financial asset through state operation—deploying financial instruments by the Chinese state, similar to the financialization of public land in the UK (Christophers 2017). This reveals the nature of state entrepreneurialism as using financial instruments built upon the city as a financial asset to achieve strategies in which the financial logic is inevitably imperative but may not occupy a central position (Wu 2023a, p. 54).

Chinese cities demonstrate some characteristics of financialized urban governance. Through investigating the process of financialization, this chapter further illustrates how a state-centered operation—a long shadow of the state—casts over financial expansion (Wu 2023a).

Figure 4.4 The Zhujiang New Town—the new central business district of Guangzhou, with clustered financial institutions.

Financing urban development is achieved through specific conduits and processes. First, there has been a housing boom. Second, there is a land-based finance. Third, the actor of state agencies (*chengtou*) facilitates land finance. Fourth, their use of non-standard financial products and *chengtou* bonds led to financialization. As a result, China's urban development is increasingly financed through financial means. The history of this financialization process also shows the deliberate deployment of financial means by the state as a strategy to cope with financial crises (such as the global financial crisis in 2008 and Chinese domestic financial pressure out of the fiscal stimulus package in 2014) and risks (as shown in shadow banking and excessive leverage by property developers). It is not a financial capitalist-initiated financialization. Specifically, housing commodification and further housing assetization mobilized households to participate in this initiative. Then, related to housing development, a series of land institutions that are quite China-specific were established. They have been created under state entrepreneurialism, using *chengtou*, land reserve centers, other LGFVs, and *chengtou* bonds. Hence, financialization is a state strategy rather than neoliberalization or financial deregulation. The state purposefully uses more market-oriented approaches to finance urban development.

Now, reflecting on financialization as a transformative urban process in late capitalism (Peck and Whiteside 2016, p. 263), financing urban development reveals that China is a capitalist-like state at the frontier of state capitalism. Its governance has been deeply interfered with by capital logic. The state is keen to experiment with the capital logic but, at the same time, constrain its operation. It has not been converted into capitalism with an ideological association (capital*ism*). In the case of financialization here, the concrete history of urban development reveals that it only partially conforms to its definition of financial expansion without the dominance of the finance sector or financial logic. Because the state represents social relations, the capitalist or financial class is absent from operating through the state apparatus. In the political economy of Chinese society, this limitation of capitalism and financial development is described as the 'market in state' (Zheng and Huang 2018) or, in the literature of Chinese urban development, known as state entrepreneurialism (Wu, 2018b, 2023b).

Notes

1 See a broad literature of land, Yeh and Wu (1996), Lin and Ho (2005), Hsing (2010), Lin (2014), Su and Tao (2017), Zhu (2019), and Qian (2022).
2 See Wu (2018b) for the comments on the concept of land finance.
3 They are known as third- and fourth-tier cities.
4 The policy bank borrows capital through special financial products from the central bank.
5 Shadow banking is outside the bank's balance sheets. This does not necessarily mean it is illegal.
6 This is from Souying Liu's public lecture, see Wu (2022c).
7 This is a component of the stimulus package.

5 Urban redevelopment

Beyond the dynamics of the growth machine

Introduction

Urban redevelopment in Western market economies has experienced a shift from urban renewal to urban regeneration (August and Walks 2017).[1] Urban renewal emphasizes the demolition of the derelict environment and the construction of public housing—post-war slum clearance under the Keynesian welfare state is one example of urban renewal. The scope of redevelopment has been greatly expanded from physical renewal to the re-use of the existing physical environment and the creation of local community vitality, which is referred to as 'urban regeneration.'

This shift from urban renewal to regeneration can be seen in the UK, which experienced several stages: in the post-war era, it was state-led housing renewal, clearance of inner-city slums, and the development of public housing (Tallon 2021). The redevelopment and construction of high-rise public housing areas led to many problems associated with the concentration of low-income people. In the 1980s, with rising urban entrepreneurialism, the redevelopment of inner urban areas and waterfront areas converted residential areas to business uses. It combined the development of office buildings with mixed uses. The latter is called 'property-led redevelopment' (Turok 1992).

Urban regeneration includes a wide range of redevelopment activities ranging from housing construction to cultural consumption. The aim is beyond housing provision and residential renovation but stresses the regeneration of the local economy, including economic restructuring, revitalizing local labor markets, developing cultural facilities, and promoting postmodern consumption. For example, the redevelopment of the Quincy Market in the inner areas of Boston has preserved traditional markets with upgraded stores, boutique shops, restaurants, and clubs (August and Walks 2017). Covent Garden in London is another example of such a project. Redevelopment often uses the provision of cultural facilities to stimulate the transformation of old urban areas, which is usually referred to as 'culture-led regeneration' (Miles and Paddison 2005).

Often, property-led redevelopment and culture-led regeneration are combined and are driven by mega-urban projects (Orueta and Fainstein 2008). The new generation of mega-urban projects did not attempt to demolish existing residential neighborhoods but built upon derelict land in former industrial and warehouse areas at the waterfront (Lehrer and Laidley 2008). In the 1990s, the new trend of 'cultural-led' or 'arts-led' regeneration constructed museums, theatres, sports stadiums, and entertainment facilities oriented toward cultural consumption (Miles and Paddison 2005; Evans 2003).

The shift toward culture-led or arts-led regeneration is due to the background of economic restructuring and inter-city competition (Tallon 2021). First, the Western economies have

DOI: 10.4324/9781003355175-6

seen economic restructuring—the decline of traditional manufacturing industries and the development of cultural or creative industries since the 1970s. Second, many cities try to attract globally important mega-events. This external factor has pushed the cities into a competition to become 'cultural cities.' For example, there is a strongly-contested competition for the 'Cultural Capital of Europe' (Miles and Paddison 2005). Cities strive to transform their inner areas with theaters, museums, art galleries, and other activities to become 'creative cities' (Evans 2009).

Urban regeneration concepts

Property-led redevelopment

Property-led redevelopment is one key concept of urban redevelopment. It is 'the assembly of finance, land, building materials, and labor to produce or improve buildings for occupation and investment purpose' (Turok 1992, p. 362). This definition includes broad changes, not just public housing renewal but also wider redevelopment processes, including attracting private sector investment into land development and, second, producing 'buildings,' often nonresidential ones, for investment and occupancy. In property-led redevelopment, property-based interests become the central element in the growth coalition of developers and local governments. They form public–private partnerships. The concept of property-led redevelopment emphasizes property interests. Urban redevelopment has given more consideration to the requirements of the financial sector than the housing needs of local residents.

Property-led redevelopment has been criticized for both its limitations and the problems it generates. The redevelopment approach tends to pay more attention to the production of properties for the market rather than the living conditions present in deprived areas. It stresses short-term improvements in the physical environment rather than long-term social progress. Such redevelopment often causes displacement, and the outcome lacks economic and social integration. The approach aims to promote economic growth, but with insufficient social and environmental objectives, there is a potential risk of overbuilding, and building office buildings in the short term can lead to financial risks. The approach is only viable when the property market goes upwards. Thus, when the property market goes down, the project becomes unfeasible.

An example of property-led redevelopment is the Dockland area and Canary Wharf in London (Carmona 2009; Brownill and O'Hara 2015). This was master-planned by SOM, a design firm headquartered in Chicago. In the 1980s, the London dockland area had become derelict because modern container ships no longer used the docks. The London Dockland Development Corporation was set up to attract investment by the UK government. A large part of the docks was given over to Canadian developer Olympia & York, who had developed a financial center in New York, so that they could start to build a second financial center in London (Fainstein 2001). At that time, the office space in the City of London was severely constrained, expensive, and unsuitable for attracting modern financial organizations. The docklands are only three miles from the City of London, and the derelict land there was abandoned and cheap.

The development was mainly for financial institutions, banks, and insurance companies, but it did also attract producer services. The development was justified by its 'trickle-down effect' which, it was felt, would benefit local communities. It was hoped that the redevelopment project would generate positive benefits such as jobs and funding for local services. The development of Canary Wharf is often described as having adopted a neoliberal approach because of the streamlining of local planning and the limited role played by the public sector. However, in reality, the government's support for public infrastructure, the DLR and the Jubilee Line, was a critical factor in the viability of this mega-urban project (Fainstein 2008; Carmona 2009).

There are many examples of mega-urban projects developed through property-led redevelopment. For example, the redevelopment of the Custom House Docks as an International Financial Services Center for Dublin (McGuirk 2000), and the La Défense in Paris became a mixed-use office, commercial, and residential development area, and a new financial center. The Guggenheim Museum in Bilbao is a typical example of culture-led regeneration. Even though relatively few people know the city of Bilbao, the iconic building of the Guggenheim Museum put it on the world map (Vicario and Martínez Monje 2003; Evans 2003). The whole area attracts tourists and generates revenue for the local government. It is often appraised as an exemplar of successful redevelopment through regenerating cultural facilities.[2]

Financialized urban redevelopment

Recent studies on urban redevelopment have paid more attention to the new role of financialization in urban redevelopment (Rutland 2010). As discussed in Chapter 3 on financialization, housing is treated as a property with exchange value during the process of financialization. It becomes a 'financial asset,' similar to the way in which land became a financial asset (Harvey 1982). In the 'financialized city,' one objective of urban redevelopment is to capture the land value. Increasingly, redevelopment is a process of 'anchoring financial capital into the city,' that is, 'the way in which capital, as it is valued in financial markets, is transformed into real capital and vice versa' (Theurillat et al. 2016, p. 1509). The literature on urban financialization emphasizes the 'importance of market finance actors' who adopt 'selective investment practice, shape urban redevelopment projects' (Guironnet et al. 2016, p. 1442).

Under the shift from entrepreneurialism to financialization, the local growth machine driven by property developers and landowners has been transformed into a debt regime dominated by bondholders, as Peck (2017a) reveals in the case of Atlantic City, NJ. Urban redevelopment under the financialized approach has to give more consideration to investors' needs:

> Accordingly, given the growing importance of investors in the ownership of the built environment, ... urban redevelopment [is] the 'outcome of power relations that originate in the circulation of investors' expectations. These expectations are met through translating market finance categories (risk, return and liquidity) into elements of the urban fabric. This bears substantial consequences for policy-making, given the current context of austerity, as municipal authorities are increasingly constrained to rely on property markets. Urban redevelopment projects are thereby increasingly shaped to provide investment assets for financial investors.
>
> (Guironnet et al. 2016, p. 1442)

The literature on urban financialization suggests that, at present, under fiscal austerity, the municipal government has limited financial resources. Therefore, it has to turn the city into an asset for financial investors. Urban redevelopment projects prioritize the needs of investors rather than urban residents. The redevelopment project is evaluated concerning a financial operation's 'risk, return and liquidity' (Guironnet et al. 2016). It is interesting within this context to see how Chinese urban redevelopment has to consider financial constraints and to what extent the urban redevelopment process in China is a process of financialization.

The application of property-led redevelopment to China

The first Chinese example of property-led redevelopment is Xintiandi in Shanghai.[3] Xintiandi is a core section of a larger redevelopment project at southern Taipingqiao Road in the former French Concession Area. This redevelopment was initiated after the Asian Financial Crisis in 1997. As a colonial legacy, this area was built in the style of terraced housing with built-in courtyards, which is known in Chinese as 'stone-portal-gates housing.' This is a style of Chinese vernacular housing, modified and combined with a Western hybrid. Housing conditions deteriorated over the years because of the lack of investment and maintenance. In the early 1990s, the Shanghai municipal government initiated urban renewal, but it could not afford to redevelop this area, which was suffering from the stagnation of the real estate market after the Asian Financial Crisis. The former Luwan district government negotiated with Shui On, a property developer from Hong Kong, to redevelop this area with offers of preferential treatment for the project. The project created a name for the place, Xintiandi, which translates as New Heaven and Earth.

The project aimed to generate profits from developing commercial, office, and residential properties in a larger nearby area while the core section of Xintiandi was carefully preserved and upgraded. The challenge for Xintiandi was over height restrictions in the area because the first Chinese Communist Party Congress had been held in one of the terraced houses. Thus, the site is subject to a preservation order. This terraced house has been turned into a museum, while another large house nearby has been renovated into a clubhouse. The architect, Benjamin Wood, who had designed the Quincy Market in Boston, was asked to develop a master plan for Xintiandi. His submitted design preserved the building style.

Without demolition, the internal spaces of several houses were merged, creating a larger floor area for restaurants, bars, nightclubs, expensive dining areas, stores, and shops. The area subsequently became a fashionable district of Shanghai. The slogan for Xintiandi is to 'Let tomorrow meet yesterday at today in Shanghai.' (Yesterday refers to the colonial era when the CCP Congress had been held there.) The project generated widespread acclaim and a professional reputation for the developer. Although there is a museum there, the primary functions are entertainment and the night-time quarter. The area surrounding Xintiandi was largely demolished, creating high-rise office buildings and luxury apartments. Its residential part, Lakeview area, is now one of the most expensive estates in Shanghai.

In terms of scale, Xintiandi is much smaller than Canary Wharf. Perhaps a more comparable example to Canary Wharf is the development of Pudong New District (Olds 2001; Wu 2003a). But Pudong is often regarded as a greenfield development as, at the time, the area was mainly agricultural fields. As will be seen in the next section, the redevelopment of old urban areas began with the renewal of alleyway housing in Shanghai. Here, we can see that property-led redevelopment has been adopted. However, there are some similarities and differences between property-led redevelopment in the UK and China (He and Wu 2005).

Regarding primary motivations and development strategies, UK and China urban redevelopments are aimed at the achievement of local economic growth, urban re-imaging, and reliance on private property development to finance urban redevelopment. These projects are located in inner urban areas or waterfront locations. The government provided favorable policies and various supports to private developers. The private developer was the principal actor in redevelopment and formed a 'pro-growth coalition' with the government. Both projects have had a social impact on the original residents who have almost all been relocated. The trickle-down effect has proven to be limited, and local communities nearby did not receive sufficient attention in urban redevelopment.

However, there are also differences between the two cases. In Shanghai, the motivation was also associated with generating capital through real estate development to finance the transformation of inner urban areas. Successful urban redevelopment is also a political achievement that helps with official promotion. In the UK, inner urban revitalization and derelict area regeneration are vital motivations. Local governments' role in urban redevelopment in the UK was minimal, while local government in China is empowered with more autonomy under the system of administrative decentralization. For example, local governments in China strongly supported property developers through the facilitation of residential relocation. Private developers play a leading role in the pro-growth coalition, as they lead the development of regeneration strategies. In China, developers had a relatively weak influence on local government. Redevelopment strategies are formulated and regulated through different tiers of government in policy intervention and financial leverage. In terms of operation mechanism, private land ownership allowed full market operation in the UK. In China, by contrast, the public ownership of land means that more non-market factors influence the redevelopment. The role of the state is paramount in this case. Local governments decide whether or not a development could happen. They often initiate development.

While there are many differences, the Xintiandi case strongly takes into consideration property interests and uses property development to facilitate urban redevelopment. This reveals that China's urban redevelopment practices are comparable with the major trend of property-led redevelopment in Western economies. However, the Chinese case presents the existence of a stronger local state, partially because state legitimacy is not subject to a local democratic political process. The state has its political intention for urban redevelopment.

In the case of Xintiandi, the Chinese government lacked capital for urban development, and the Asian Financial Crisis weakened the real estate market. Thus, the government gave greater support to private developers to regenerate inner urban neighborhoods. During this period, the private sector developer led the project's operation. However, the private sector developer and the market have been used as a means to achieve political aspirations and the needs of the state in terms of promoting officials or revenue generation for the local state.

With regard to state sovereignty, there may be less difference between China and other market economies. Even in London, the local growth machine dynamics might not dominate in the agenda of urban redevelopment as local councils step up their companies to deliver affordable housing or subsidize the public sector operations (Beswick and Penny 2018). This change in financing approaches should be understood within its 'financial context' (Christophers 2019). The mega-urban projects extract value to finance affordable housing delivery (Robinson and Attuyer 2021). Similarly, in suburban areas of London, the development has multiple objectives other than financialization (Ferm and Raco 2024).

Although the objectives of land value capture and social development are intertwined, the local state has an agenda independent from profit-making and financialization in urban redevelopment. Seen in this way, property-led redevelopment should not be regarded as finance-driven regeneration. Developing financial centers and business districts in China often represents a state strategy for a larger region. For example, in Shenzhen, the development of Qianhai is considered to integrate Hong Kong and Macau into the whole region under the Greater Bay (Figure 5.1). In this context, property-led redevelopment aims to achieve a political end (see Robinson et al. 2022 for a comparison of London, Shanghai, and Johannesburg). This means that we should further reflect on the concept of property-led redevelopment.

Figure 5.1 The development of a new financial center—Qianhai—in Shenzhen as part of the Greater Bay strategy.

Property-led or developer-led redevelopment

As will be seen later, although China also uses 'property-led redevelopment,' this does not mean that property developers led the regeneration agenda. We need to clarify the concept, which is quite problematic because it is compounded from two different meanings: first, it refers to the change in land use from residential to non-residential uses (for example, redevelopment projects include the component of office and commercial property development). Therefore, the redevelopment project goes beyond residential development, including office and commercial development. Second, the concept implies the role of actors other than the state in earlier public housing renewal—property developers.

To understand this concept properly, we must examine the state and market relationship because property redevelopment often involves private-sector developers. This approach to combining property development with public housing renewal led to a more significant role for the private sector in urban redevelopment. The rising role of developers in urban redevelopment is driven by the political economy of neoliberalization, or entrepreneurial governance, as discussed in Chapter 1 on state entrepreneurialism. One objective of entrepreneurial governance is to attract investors. As stated earlier, property-led redevelopment attracted investors, such as Olympic & York in Canary Wharf in the late 1980s. The mega-urban projects of property-led redevelopment are often initiated by the developer rather than the state, which is read as evidence of entrepreneurial governance. By contrast, mega-urban projects in China are less likely to be an initiative of the private sector.

Figure 5.2 The new popular commercial complex at Yushenli historical neighborhood near Suzhou Creek in Shanghai.

Yet property-led redevelopment may not always be equivalent to developer-led redevelopment. In some cases, developers do indeed play a critical role. However, even though developers are important in other places such as China, they are often introduced or utilized as a market instrument to achieve government intention. Thus, empirically, more research is needed to understand the role of developers and carefully examine them in urban redevelopment. Their role needs to be understood within the context of a political economy, particularly urban politics. If the context is different from that under neoliberalism, then the role of developers may be subject to more substantial public intervention. For example, in Europe, urban development, in general, is subject to a stronger regime of city planning than in the USA (Haila 2016; also see Robinson et al. 2022 about territorial politics). As shown in the Xintiandi case, Chinese urban redevelopment differs from that of the UK in that the local government in China still plays a significant role, for example, facilitating residential relocation and vacating the development site. However, in recent Chinese micro-redevelopment practices, we witness a visible role of the central government and associated discourses. To what extent does the 'property-led redevelopment' concept still help illuminate the reality of Chinese urban redevelopment? It seems that the model of Xintiandi continues to be replicated in China (Figure 5.2). Thus, we will further interrogate the concept of property-led redevelopment in this chapter.

To summarize the concepts involved in urban regeneration, we first focus on 'property-led redevelopment.' The shift from slum clearance and physical improvement to wider local economic development has been accompanied by the application of market operations—the decline of the Keynesian welfare state and rising neoliberalism. Property-led redevelopment is related to rising entrepreneurial governance. China has seen a similar trend. However, state sovereignty is not abandoned. We need to understand state politics to comprehend urban redevelopment dynamics. Thinking along this line, fiscal austerity and financial crisis mean that more consideration has been given to financial interests. However, because of territorialized politics or the territorialization of the state, financialization sets the context rather than the end because the state is not singly driven by the process of financialization or pursuing financialization. Because of fiscal austerity, local governments have to compromise their development goals. This does not mean that financialization will become a policy goal. Financialization, in various cases, is not the determinant for urban redevelopment.

Urban redevelopment has recently become a priority of state politics in China. To understand property-led redevelopment alongside the recent trend of entrepreneurial governance and financialization, we need to review the historical stages of urban redevelopment to see how national politics influence redevelopment in response to the changing political economy. In other words, property-led redevelopment should be viewed as a tactic of the state to cope with the threat to structural coherence of capital accumulation (Wu 2017; Deng 2023).

Changing practices of urban redevelopment

Historical stages[4]

China's urban redevelopment started as the process of 'housing rehabilitation' in the 1980s. The initial objective was relatively straightforward: to improve the living conditions of dilapidated housing in inner urban neighborhoods. Most residents remained in their houses after housing rehabilitation. However, housing rehabilitation was expensive for public finance, and municipal governments could not sustain such programs. Therefore, urban redevelopment only proceeded on a minimal scale.

Large-scale urban renewal began to occur in the 1990s, when housing and land markets were established. During this period Shanghai launched a '365 housing renewal program' to redevelop 3.65 million square meters of old housing. Similarly, Guangzhou used the opportunity of metro station construction to initiate urban renewal. With the establishment of real estate markets, the nature of urban renewal changed from improving the condition of existing housing to generating profits and revenues from land development. Property-led redevelopment has sped up urban demolition and the transformation of Chinese cities since the Asian Financial Crisis in 1997.

Starting in the 2000s, urban redevelopment has been driven by ambitious mega-urban projects to build world-class cities. Urban demolition became a salient feature, leading to frequent social contests (Shao 2013; Shin 2016). The state plays a significant role in urban redevelopment, although the real estate market is utilized to finance redevelopment projects. This redevelopment approach reveals the feature of 'planning centrality, market instruments' under state entrepreneurialism. The logic of redevelopment includes revenue generation and building world-class cities. This differs from gentrification driven by middle-class household consumption and real-estate profit-making (Wu 2016b; Wu, 2020b). In Shanghai, the demolition of urban villages aimed to create office buildings, business and technological parks, and new green spaces. Large-scale retail gentrification and culture-led regeneration selectively preserved original architectural styles, but converted residential uses into shopping districts, creative industry parks, and tourist and entertainment quarters.

Since the start of the 2010s, the cost of land acquisition has significantly increased. This makes it more and more difficult to relocate residents, who demand more compensation. In response to contests and the threat of social instability, Guangzhou began to explore small-scale redevelopment, known as 'incremental regeneration' (or micro-redevelopment, *wei-gaizao*). Since 2015, small-scale redevelopment has been experimented with in many Chinese cities. Numerous micro-redevelopment projects are mixed-use developments, transforming original residential uses into restaurants, museums, boutique stores, and tourist attractions. Such small-scale redevelopment preserves the physical appearance of historical buildings but leads to significant changes in land uses and functions and displacement.

Since the late 2010s, the pace of urban redevelopment has been accelerated. Redevelopment has become a policy priority, as urban expansion and the development of green fields are subject to stricter regulation under environmental concerns. The real estate boom stimulated by the fiscal stimulus package since the Global Financial Crisis in 2008 has increased the financial leverage of real estate developers and unsold housing in smaller cities. The central government launched large-scale dilapidated housing renewal to maintain financial stability and further solve the problem of local government debt. Through the China Development Bank, the renewal programs provided monetary compensation to residents so that they could buy new commodity housing.

Along with the broad process of housing financialization, 'shanty-town' renewal resorted to financial means (He et al. 2020; Wu et al. 2020). The large-scale shanty-town renewal programs also extended to rehabilitate 'old and dilapidated estates.' However, these estates, created initially as planned 'micro-districts' with multistorey buildings, deteriorated over time and lacked parking spaces and elevators. It proved impractical to demolish them. Instead, they have been improved through incremental refurbishment.

In sum, China's history of urban redevelopment experiences several stages: (1) limited housing rehabilitation in the 1980s and enlarged renewal programs in the 1990s; (2) mega-urban redevelopment projects through demolition and property-led, mixed-use developments in the 2000s and 2010s; (3) small-scale or micro-redevelopment experiments with increasing financial means since the mid-2010s.

The history of urban redevelopment in China reveals several features. First, the state is heavily involved in urban redevelopment, directly through state-owned development corporations and financial means and indirectly with state-led redevelopment programs.

Second, the state has various or multiple motivations, contingent upon historical moments and conditions, including dilapidated housing upgrading, revenue generation through collaboration with private-sector developers, promoting economic restructuring toward post-industrial economies and global cities, maintaining financial stabilities, and coping with the impacts of the global financial crisis.

Third, the changes in inner urban areas through urban redevelopments are mainly driven by supply-side initiatives rather than by demand for lifestyles and amenities. This does not mean the preferable living conditions in the central area due to better public services than in the new towns on the periphery are irrelevant. However, redevelopment is associated with China's overall political and economic changes, such as transforming the central areas.

Therefore, redevelopment differs from classic gentrification—the replacement of original working-class residents in their neighborhoods by middle-class newcomers (Glass, 1964). The definition of gentrification has greatly expanded and now includes a variety of forms (Lees et al. 2016). The history of redevelopment in China suggests that urban redevelopment might

Figure 5.3 The redevelopment project along the Erhai Lake in the city of Dali attracts tourists and new residents taking wedding photos.

be seen as a form of 'state-led gentrification'—state-orchestrated neighborhood changes—if we wish to retain the term 'gentrification' (Wu 2022a).[5]

Redevelopment projects often combine state environmental protection programs with consumption-based urban development. For example, in Dali, a popular tourist city, guest-house development driven by the new rural lifestyle and the state mandate for protecting Erhai Lake are key forces for regenerating old areas and constructing the ecological corridor along the lake (Figure 5.3).

Urban redevelopment becoming a policy priority

Chinese cities have experienced rapid expansion since the land reform and the adoption of land-driven development. However, land acquisition has become more costly in recent years. As a national strategy, the state protects agricultural land and restrains suburban land development. As Chinese cities grow larger, the regeneration of inner and central urban areas becomes a policy priority. Because of social contest over demolition, wholesale demolition has been replaced by small and incremental redevelopment. More research is still needed to shed light on the shift toward a redevelopment approach.

The history of urban redevelopment reveals several features. First, the state is heavily involved in the redevelopment process and plays a prominent role. Second, there are various motivations and policy objectives, including dilapidated housing renewal, land revenue generation, real estate speculation, post-industrial economic restructuring, global city building, and the creation of development opportunities to sustain economic growth. The motivations are mixed. It is thus essential to investigate the actual motivation of the state in urban redevelopment. Why does the state initiate urban redevelopment programs to formalize urban spaces and adopt small-scale regeneration? The following two sections examine urban redevelopment in villages and inner urban areas.

Urban village redevelopment

Urban villages as informal settlements for migrants

In this section, we will examine urban villages. These former rural villages either inside or near the city have gradually been encroached upon by urban expansion. They evolved into informal settlements of rental housing. Some villages also developed production activities such as small workshops.

First, we will examine the main features of urban villages. Then, we will discuss different outcomes of urban village redevelopment. The difference largely depends on the timing of redevelopment as well as the strength of the local community of villagers. Urban villages change all the time. Large-scale demolition has been a primary form of redevelopment, as the building quality of farmers' houses is generally low. Urban village redevelopment intends to remove informal development or self-built housing (Wu et al., 2013). We will introduce new redevelopment practices such as 'three-old redevelopment' (the redevelopment of old villages, neighborhoods, and industrial areas).

Chinese rural areas experienced significant changes. In rural–urban migration and urbanization, young people left their villages to work in the cities. Some villages became empty, with those remaining being the elderly and children. Some rural villages managed to change their functions during urbanization, for example, by introducing 'creative industries.' For instance, Xiaba village in Dongguan developed oil-painting craft workshops for Western and Chinese new middle-class consumers. Others, like Xiaozhoucun in Guangzhou (Qian et al., 2013) and Songzhuang in Beijing, experienced upgrading of famers' houses into artists' studios (Zhang 2019). Through these developments, both the village environment and facilities have improved. Small restaurants, cafeterias, and pubs serve the tourists and new residents. Some villages see a concentration of 'artists,' but many are workers in the production line, for example, in the production of replica oil paintings in Shenzhen (Wang and Li 2017).

However, in large cities, former rural villages have been encroached on and converted into 'urban villages.' Some accommodated small workshops. Many are migrant enclaves with rental housing built by the villagers. In Shanghai, the former village of Gaojiabang was hidden behind a clean street in the prosperous Xuhui district in central areas. Just a single small gate led to a large informal settlement. Outside the village, the roads are well-managed and clean. The residents' committee also managed the village as it had been converted into an urban area during land requisition by state-owned enterprises in the early 1990s. The place became semi-urbanized, as seen from a security guard at the gate. However, the living conditions inside the village were deplorable and very congested due to the subdivision of rental rooms for multiple occupants. Although there are a lot of small grocery stores, restaurants, and barbers, the public facilities are inferior, with few indoor toilets. Most migrants living in urban villages have jobs, although these are generally in low-paid social services. The urban village provided convenient accommodation to migrant workers with good connections to the outside world. Thus, urban villages are not ghettos isolated from the rest of the city as ethnic enclaves. Residents are not excluded from the outside world, but they are poor.

In contrast to deteriorating urban villages in Shanghai, urban villages in Guangzhou and other cities in southern China have better conditions, mostly experienced entirely new self-built by villagers. Former farmers' houses were redeveloped into multiple floors, some even with elevators. The quality of buildings is much better, and they provide purposely built accommodations for migrants. In Zhucun (the Pearl Village) in Guangzhou, some buildings present a very odd shape with large upper floors and a smaller base on the ground. This is because every house has been individually rebuilt up to the boundary of land plots to use the maximum space. The upper floors even extended into more airspace of alleyways. So, buildings are very close to each other.

Villagers thus became de facto property developers (Sa and Haila 2023).[6] Although each building is well built, the overall living conditions are undesirable, and the public or communal space lacks maintenance, with litter, rubbish, and dirty ponds. In addition to rental units, some village houses are converted into production and warehouse spaces. A few are developed on a large scale; for example, the village outside Sun Yat-Sen University was developed into a wholesale clothing market. Small workshops are moved in and clustered around the market. The place thus became a mixed-use area with a wholesale market and small workshops and living space for workers. The development of industrial uses inside urban villages is due to a rather lax mode of management and a stronger local community based on the power of lineage. In Shanghai, near the Jiuxing wholesale market for DIY and building materials, small workshops were created in former villages. However, in contrast to the wholesale market in Guangzhou, urban villages in Shanghai were later demolished because of the strong government capacity.

From these examples, we can see that urban villages are very dynamic. The building qualities vary. With stronger government control in Shanghai, the urban village maintained their original farmer's houses with appended and subdivided rental rooms. However, urban villages have seen large-scale self-construction in South China, where the local community is stronger with the tradition of lineages. Individual landlords rebuilt their houses for rental. So, urban villages are evolving and embedded in urban life and production. Some even became large wholesale markets.

Village Redevelopment

Over the years, the living conditions of urban villages deteriorated despite the strong demand for affordable housing. Ironically, the lack of redevelopment and the poorer environment are due to state control rather than state incapacities. Villagers cannot demolish their farmer's houses to construct new residential buildings; they can only append additional floors and subdivide their rooms for rental. The weak village governance is also due to the dismantling of village organizations during state-led industrialization when capable villagers were recruited as workers of the state-owned factories.

Because of state land requisition, land outside the villages became state-owned, and villages owned residual land. They were left out of redevelopment because of smaller plots and high population density. Because of the poor building qualities of the original farmer's houses, the foundation of the buildings could not sustain more additional floors. The informal building activities are unsafe. Many villages also face future demolition. Thus, villagers wait for compensation rather than self-improvement. They only managed to append as much additional space as possible to maximize the rent.

The redevelopment of urban villages can often take a demolition and rebuilding approach. For example, Gaojiabang in Shanghai has been demolished and rebuilt as part of the Caohejin Development Zone. The informal residential land use has been transformed into the office park for Shanghai's modern producer service economy. The village redevelopment is associated with the process of industrial restructuring. In Guangzhou, for example, Pazhou was demolished in order to allow for the development of the Guangzhou Convention Centre. Although local communities in Guangzhou have a larger self-organization capacity and have been capable of improving rental housing conditions to a greater extent than urban villages in Shanghai, they face a similar fate of demolition initiated by the state. Urban demolition is widespread in Chinese cities, and it has attracted extensive research attention and critique.

China is an outlier regarding state capacity and the scale of redevelopment and demolition (Hamnett 2020). If we wish to consider Gaojiabang's redevelopment gentrification, we need to provide two qualifications: this development converted residential uses into office buildings of an industrial zone rather than residential changes; and the redevelopment is state-led. Hence, this would significantly expand the meaning of gentrification to such an extent that it is almost equivalent to displacement, which occurred in this redevelopment process.

In this sense, village redevelopment is not a process of classic gentrification but rather one of an extended form of 'gentrification' (meaning state-driven redevelopment). Considering this new feature of state-led development strategies, gentrification is referred to as 'a global urbanism' (Smith 2002). This redevelopment approach is qualitatively different from the classic notion of gentrification (Hamnett 2021). The redevelopment resembles building world-class cities in other places. In other countries like India, this type of redevelopment aims to eliminate informalities. It is driven by the 'gentrified state' (rather than individual households) through the state's speculative urban development programs (Goldman 2021), with an objective of aestheticization (Ghertner 2015).

In some specific circumstances, the state may allow villagers to play a more significant role in urban village redevelopment. For example, the Liede village in Guangzhou was allowed to use developers to redevelop high-rise buildings with very high plot ratios, and they received generous compensation. This new form of project replaced previous state-led redevelopment. It differs from the example of Tangjialing because the state did not acquire the village land for real estate projects. In the case of Liede, it was more village-led but still took a form of real estate development. After redevelopment, villagers happily organized huge banquets to celebrate their new homes. Having several apartments from compensation, they rent out to new tenants who are office workers in the nearby central business district rather than to the previous migrants.

In contrast, Liede is an exception because of an overall increase in density, with very limited land contributing to public uses. This is because Guangzhou was pressured to redevelop Liede, and the village governance was well organized. The Shenzhen case demonstrates even more apparent features of state entrepreneurialism than the bottom-up regeneration which is happening in Liede. In Shenzhen, new practices exist to develop long-term social rental housing (Li et al. 2021). The aim is to ask the villages to provide affordable housing to key workers through land adjustment and redevelopment. In these cases, we can see urban redevelopment has multiple objectives. In Shanghai, it was associated with economic upgrading. In Shenzhen, the redevelopment also helps expand affordable rental housing.

In Beijing, villages rented land to small developers in order to develop purposely built residential compounds, targeting better-off migrants. Some villages developed and sold 'small property-right houses' to sell to urban residents (Liu et al. 2012; He et al. 2019). The quality of this purposely built rental housing is relatively good. It is generally clean and affordable to migrant workers. The careful design ensures natural lighting and good ventilation. Most have internet connections, and some compounds provide small shuttle buses to connect to the stations. Developers have tried to tailor their products to the market demand. Some are called 'white collar apartments,' indicating that their customers are from a better socioeconomic status than migrant industrial workers. For Tangjialing, many renters work for IT industries.

Informal rental housing development in peri-urban villages like Tangjialing has attracted media attention. The book *Ant Tribe* describes the migrants and university graduates working in IT industries near the software park of Zhongguancun Science Park in Beijing. IT industries have attracted more migrants since 2000. These migrants often called themselves 'IT migrants' or, as the book title suggests, the 'ant tribe,' that is, without stable accommodation. The village also leased the collectively owned land to small builders to operate large residential compounds of customarily built rental housing—which became known as 'white-collar apartments.' Although since 2005, the villagers' committee has tightened the control over land to slow down self-rebuilt activities, construction is still widespread. With the influx of migrants, Tangjialing was also described as a chaotic and dirty place. This pushed the local government to demolish Tangjialing in 2010. However, the Tangjialing New Town project builds affordable housing for Beijing residents (Figure 5.4). IT migrants cannot afford the increased rents and they do not qualify for affordable housing. So, urban village redevelopment is largely a removal of informal rental housing, even though the informal rental housing provided affordable accommodation to local workers.

Figure 5.4 Tangjialing New Town, a large affordable housing estate built by the Beijing Municipal Government.

From these cases of village redevelopment, we can see that existing landowners try to capture land rents by refurbishing their residential buildings. However, the redevelopment initiated by the state is intended to remove the informalities created by self-building. The redevelopment process also demonstrates some flexibility. The original village sites were developed into a new town, an office park, or large residential areas of multiple high-rise buildings. The high plot ratio guarantees an investment return. The state has to be more flexible about density control in order to make redevelopment happen.

Rethinking urban village redevelopment beyond collectivism

Urban village redevelopment is not a community-led redevelopment. It is a form of redevelopment under 'state entrepreneurialism,' which gives incentives to society and uses market approaches for redevelopment (Wu 2018b). This strong state role may suggest China is an outlier in informal settlement redevelopment (Hamnett 2021) since this type of redevelopment is still state-managed, and subject to the state's guidance. All redevelopment projects must be qualified and approved under the government's 'inventory of redevelopment,' specifically the *sanjiu* office.

Meanwhile, local governance is an important factor that has different outcomes. Stronger communities may result in a better deal for villager owners. But still, for the state, redevelopment is more 'crisis management'—in the case of *sanjiu* to cope with the impact of the global financial crisis on the export-oriented manufacturing industries in the Pearl River Delta or problem-solving like Beijing's housing affordability—than purely profit-driven.

Liede was redeveloped because it is located near the site of the Asian Games and the Guangzhou new CBD (Figure 5.5).[7] The municipal government needed to develop a new image for Guangzhou—such an image building is a political consideration and achievement. Seen in a larger context, Liede is an experiment of a new development approach (Lin 2015).[8] The redevelopment strategy is a 'fix' for Guangdong's capital accumulation problem. It is a policy exception negotiated by the former party secretary, Wang Yang, with the Ministry of Land Resources. As Guangdong faced limited land resources, the secretary asked for special treatment to allow a more 'bottom-up' approach to redeveloping existing villages, old estates, and industrial land without going through formal state land acquisition through auctions.

The bottom-up or village-led approach means that as long as the project is qualified, the villagers can find developers and organize redevelopment activities, expanding the land supply for overall economic growth and urban development. This was urgent as the Global Finance Crisis hit hard on the export-oriented economy in 2008. In the case of *sanjiu*, the state made a generous deal and did not maximize land value capture. It forgoes land revenue to local communities. Nevertheless, although the state does not pursue land value maximization, the redevelopment adopts and reinforces wholesale demolition because the redevelopment requires large-scale real estate development (rather than self- or new middle-class gentrification) to finance these projects.

The new form of *sanjiu* redevelopment is coordinated by the village. It has a comprehensive plan for the whole village rather than the development organized by individual households on their land plots. For example, in Shenzhen, redevelopment divides the city into urban regeneration units. Land ownership is adjusted, and landowners are asked to contribute land for public facilities such as public spaces and schools and receive additional development densities in other places on the site. On average, 30 percent of residential land has been adjusted for public use, while the owners received compensation through development density increases. The objective is to improve the overall living environment.

Figure 5.5 The Liede Village area near Guangzhou's new central business district.

Thinking for the Global South: informality permitted by the state

What are the lessons of the redevelopment experience for the Global South? On the positive side, China seems to be able to redevelop informal settlements, compared to state incapacities and prolonged slum conditions in other cities of the Global South.[9] Urban village redevelopment in China is associated with dynamic economic growth and broad urban transformation. The capacity of state-centered governance is a key factor in informal settlement regeneration. The strong state uses market instruments to achieve its redevelopment task, in contrast to the weak capacity of the state in the Global South.

On the other hand, China cannot entirely solve the problem of low-quality housing for the poor. Although redevelopment demonstrates the effectiveness of market development and villagers' involvement, the excessive constraints from state regulation hinder sustainable regeneration. The redevelopment is far from a community-led self-upgrading. The enforced wholesale demolition approach to village redevelopment displaced migrants and low-income renters to other places, which created new urban villages elsewhere. China also sees a prolonged replication of the low-quality housing problem. The persistence of informal housing development in peri-urban areas is due to the marginal social citizenship status of rural migrants in Chinese cities and their lack of participation in decision-making. Their right to the city is constrained (Qian and He 2012), and they wait for the state to bestow necessary conditions and benefits, even though they managed to explore a flexible and informal way of finding accommodation based on their own agencies (Smart 2018). The social agency is

Figure 5.6 The Nantou Village has become a new tourist and consumption destination in Shenzhen.

subject to the structural constraint imposed by state entrepreneurialism—as migrants face restrictions from the state and market.

Urban villages are informal developments created by villagers' self-building activities. Their redevelopment is often initiated and organized by the state in order to eliminate informalities. There are variegated outcomes, depending upon the strength of local communities. Some villages see continuing deteriorated housing conditions and a poor living environment because village governance is relatively weak. Some villages with strong self-governance capacities based on lineage manage to conduct extensive rebuilding under the flexibility given by the state. Intense social contestation on demolition may lead to more generous compensation and policy flexibility. The degree of flexibility is also dependent upon some historical and local circumstances. When the government needs to rebuild the informal development area for some mega-events, the policy becomes more flexible and gives more interest to villagers. The state also permits the more extensive use of the market development approach, as shown in *sanjiu* redevelopment in Guangdong. Through redevelopment, some villages are turned into popular recreational, tourist, and shopping places (Figure 5.6)

Overall, village redevelopment is not a self-organized process. It has the visible intervention of the state. Thus, in the informal settlement of Chinese cities, we do not see self-upgraded gentrification under neoliberal governance (Wu 2020b; Zhan 2021). Village redevelopment reflects state entrepreneurialism, either using market actors—developers by the state—or permitting villagers to develop through the market.

Preservation, improvement, and micro-redevelopment

The small-scale redevelopment began with an experiment in Guangzhou. Like Xintiandi in Shanghai in the 2000s, Yongqingfang is allegedly the pilot project of the new model in the 2010s (Wu et al. 2022b).[10] The scale of this pilot redevelopment is smaller. The historic neighborhood is located in central Guangzhou and has arcade-style buildings along the street (Figure 5.7). The building style is regarded as a heritage of southern China. Previously, the redevelopment of Enning Road, a larger area within which Yongqingfang is located, demolished the old buildings. The large-scale redevelopment project caused social contestation, including the appeal to the central government. The redevelopment project did not proceed as expected. Many sites were left as ruins or unfinished constructions.

One important change for small-scale redevelopment projects is the changing government motivation. The local government does not wish for them to become real estate projects, although the projects use real estate development to finance redevelopment. These projects are initially perceived as property development projects not intended to capture land values but rather to achieve social objectives of regeneration, heritage preservation, and job creation. Some projects claim to 'make people happier,' and some attempt to preserve and 'appreciate the local heritage and culture,' triggering people's nostalgia. These objectives can be seen as state propaganda; at the same time, however, they provide policy guidance. While this could

Figure 5.7 Yongqingfang redevelopment project represents a new model of 'incremental regeneration' in Guangzhou.

be just discourse, the guiding principle has changed. It is intriguing to ask why urban redevelopment has changed its objective.

Yongqingfang, which was artificially created for a project, consists of several plots of largely public housing with some private houses. Many residents have already been relocated during the demolition phase. The district government controlled the empty public housing, making land conversion possible. As such, it was easier to redevelop this site. The neighborhood was thus selected for an experiment on micro-redevelopment.

A large proportion of the site was designed as a co-working space for creative industries. Other uses include education and youth rental housing, which have not materialized to date. Some houses were converted into small boutiques, stores, restaurants, and craftwork shops along the alleyways. The area created for co-working space was also former public housing. Small museums, such as Bruce Lee's Ancestor House, were also created. Compared with Xintiandi, these shops are less expensive and smaller. There are few expensive diners or clubs.

Vanke, a larger domestic enterprise, was developed based on a system of build–operate–transfer (BOT). Land ownership has not been transferred from the state to the developer, who remains only the asset manager for commercial and office rentals. The style of residential buildings is preserved, so it is also regarded as heritage preservation. The developer raised development capital, but gained rental income from non-residential uses. It plays the role of the property management company. Vanke can anticipate an increasing rental income when the place becomes more popular. The profitability is based on a long-term lease. Nevertheless, it is challenging to cover the redevelopment costs. The redevelopment has so far been relatively modest, and it has been reported that Vanke recovered the cost of redevelopment within five years, well ahead of the original plan. The rental has increased due to its popularity because Yongqingfang has become well known since the visit of President Xi Jinping in 2018, when he endorsed this historical preservation model.

The project is presented as an exemplar of inner urban regeneration because it preserves the historical building style. It attracted many visitors, including officials from other places, to learn about the redevelopment approach. Through redevelopment, the conditions of buildings have been improved. The neighborhood has changed from a purely residential to a mixed-use area. Only a couple of households remain and live in the neighborhood, which has now become largely a tourist site and an office workspace.

It is difficult to refer to this redevelopment project as culture-led regeneration because of the disappearing neighborhood and local culture, despite the preservation of physical building styles. While the initial motivation was to regenerate old neighborhoods, this objective was not achieved, as often occurs in urban redevelopment projects. The claim for 'small-scale' redevelopment is still questionable, as the change in land use is significant—from residential neighborhoods to tourist and office space. The so-called small-scale and incremental redevelopment generates substantial changes in urban landscapes. The area also sees an upgrading from traditional shopping streets to a trendy tourist place. Because of commercial and office space development, the project might be called property-led redevelopment, commercial gentrification, or retail gentrification (Hubbard 2018). The latter refers to transforming from local shopping streets to commercial spaces for middle-class consumption (Gonzalez 2020). It shows an upscale business and changes in land use. However, these concepts, in their current meaning in the literature, do not fully describe the nature of these micro-redevelopments in China.

Micro-redevelopment has also been extended to redeveloping relatively modern residential estates. These areas have deteriorated over the years since their development before the 2000s and became known as 'old and dilapidated estates' (Wu 2018a). The motivations for regeneration are mixed. Most did not aim to demolish them or to become a real estate project. Social governance is stressed. Urban redevelopment is an experiment for a new social governance.

Some promote the co-governance agenda, encouraging public participation and social mobilization (Li 2022). The physical deterioration needs a degree of public intervention, as many five to six multi-story houses without elevators and their residents have become older. The neighborhood lacks parking and recreation spaces. There has been an initiative to regenerate these dilapidated estates. There are different explanations for the initiative. One perspective is to situate this type of redevelopment in capital circulation and capital accumulation. After the Global Financial Crisis in 2008, the Chinese government began a new round of financialization, as discussed in Chapter 3. In 2015, the urban redevelopment initiative was raised to support real estate development to mitigate financial risk (He et al. 2020). Another perspective stresses the multiple logics of the state. In this case, the dilapidated housing renewal, just as its operation in the 1980s, has both a social and a political objective. Under President Xi Jinping's leadership, one of these objectives is 'common prosperity' and a happier society.

For example, one neighborhood in Huajin in Wuhan has become an exemplar for co-development and co-governance (Wu et al. 2022b).[11] The redevelopment project was advised and designed by researchers from Wuhan University and funded by the government. The project stresses grassroots participation. The party secretary has coordinated the project in the residential community, strengthening the party members' leadership. When thinking about politicians in Western democratic political regimes, who may also come from businesses and serve as community builders, it is interesting to note that party members in China are encouraged to serve as community builders in their own way.

The experience in Huajin is very different from that of Yongqingfang in terms of land use change because the former does not involve residential relocation or conversion from residential to commercial uses. However, there is a similar underlying logic behind these micro-redevelopments. They all present the use of social and market instruments to achieve some state objective. There is only one logic to promote the centrality of the state agenda, albeit with diverse pathways and forms. Both encourage social participation, though Huajin involves more explicitly the capacity of party members.

Social participation is thus an ideology, a policy discourse, and a requirement for physical regeneration. Down to the bottom level, there is a lot of practical coordination and reconciling needs for neighborhood conflicts, such as different preferences and elevator installation options. Community volunteers are called upon. Thus, the redevelopment also experimented with 'neighborhood planning.' Some volunteers are given the title of 'neighborhood planners.' This is alongside the transformation of city planning from a top-down approach envisioned by elite professionals to a more bottom-up approach involving neighborhood participation.

Currently, these projects are funded by the government, but in the long run, the cost needs to be recovered from their users—residents. In reality, residents cannot shape the redevelopment agenda, or at best they can only do so in terms of short-term, concrete, and physical construction rather than having any decision power for agenda setting. It will be an interesting direction for future research to observe whether or not China will see a bottom-up process of democratic movement from its neighborhood politics. The current neighborhood governance agenda is 'co-governance'—encouraging social participation in state-formulated agenda.

Beyond growth machine dynamics[12]

State intentionalities

The state in China has diverse interests, intentions, and motivations to promote urban redevelopment. Real estate interests do not necessarily bind these motivations. The state has its own territorial logic to maintain the structural coherence of its political economy, including maintaining political stability, presenting itself as a benevolent guardian of society, and containing economic crises. It is, therefore, important to understand the contextual particularity regarding the extent and forms of state politics (Wu et al. 2022b).

From the history of urban redevelopment, political considerations other than profiting from land are salient in the redevelopment programs. These considerations are often presented as 'national political mandates' that guide local development practices, even though the local government has some discretion in actual implementation or even twists the policy toward its benefits. In village redevelopment programs, one crucial mandate is national food security. Farmland is protected out of this consideration. Urban redevelopment encourages using existing land more efficiently rather than occupying more farmland. This consideration includes environmental sustainability and social stability (Lin 2015; Zhang et al. 2022).

The central government initiated a series of campaigns that guided local development. These campaigns exert top-down general guidelines without detailed implementation methods. They present as discourses, principles, or even slogans. However, local government officials must understand the importance of, and adhere to, the spirit and redlines. Some mandates are particularly influential in urban redevelopment.

First, *rural vitalization* is broadly related to countryside preservation and economic upgrading. In contrast to treating village redevelopment as real estate projects, this mandate aims to enhance the overall vitality of the rural economy, for example, by promoting tourism. The policy, often associated with anti-poverty programs, aims to improve the rural living environment. Rural landscape preservation is emphasized through the 'Beautiful China' program. This justifies preserving rather than demolishing rural vernacular housing. However, during the actual implementation, preservation becomes quite selective. Farmhouses are turning into guesthouses and small hotels, and the rural landscape is often artificially created for tourism (Su 2015).

Second, *heritage preservation* has been given more significant consideration. Preserving traditional building styles may be a real estate tactic or an outcome of gentrification; for example, the 'new rich' renovate traditional courtyard housing and move into alleyway (*hutong*) areas. In Beijing and many other Chinese cities, the cultural protection districts for heritage protection are designated as government cultural preservation policy. The media, planning and design professionals, university researchers, NGOs, and residents recognized the value of heritage, forcing the government to change the demolition approach to heritage preservation (Figure 5.8).

Third, *community participation* aims to emphasize social governance. This is a more explicit political consideration. Besides achieving a greater urban order and environmental aesthetics by removing informality and disorder, the government encourages community participation, making residents happier. However, many residents do not bother to participate in the activities organized by the government, neighborhood organizations, or local planners. While traditional urban renewal emphasizes the rehabilitation of the built environment, the purpose of community participation adds the dimension of social well-being. It is also hoped that participation enhances the cohesion of the residential community, mutual surveillance, and social

Figure 5.8 The incremental redevelopment approach now pays more attention to culture and heritage, as shown in a neighborhood of Bantang Wuyue in Guangzhou.

stability. Practically, providing public facilities in dilapidated neighborhoods often requires mediating the conflicting interests of residents. For example, residents on the different floors of the multistory building have different preferences for installing an elevator, and hence, before incremental redevelopment can proceed, community involvement is needed.

In short, these mandates transformed urban redevelopment practices from the earlier property-led approach to new micro-redevelopment with decisive state intervention. Although property developers still conduct many rehabilitation projects as real estate development, incremental urban redevelopment is subject to broader state politics and national political mandates. This means redevelopment programs are initiated from political considerations rather than pure profit calculation. At the same time, the local government must consider the real estate market's financial feasibility and profitability. Local redevelopment practices must align with central government policies, meaning these projects are often branded with a different audience—to demonstrate their achievement under these mandates.

Implementing through the multi-scalar state

Recognizing the importance of 'national political mandates,' how does the multi-scalar state operationalize these policies in local redevelopment practices? First and foremost, the mandates reflect a shift of political ideology and state ethos under the changing political economy and geopolitics. The new redevelopment projects depart from the earlier property-led redevelopment under entrepreneurial governance. Although real estate development is an important driving force for economic growth, rampant demolition and redevelopment aggravate the capital accumulation crisis and threaten structural coherence. Because property-led redevelopment generated severe social contest, instability, local debt, financial risks, and environmental degradation, the new state ethos is to curtail market operation to maintain political stability.

The mandates stress village improvement rather than relocation and dispossession, job creation in creative industries, and heritage protection. For some projects, like Huajin in Wuhan, the government directly funded the project to refurbish existing housing and facilities to promote neighborhood cohesion and social mobilization. This is not a property-led redevelopment for generating land revenue. The underlying motivations of these redevelopment projects are beyond market-driven real estate and entrepreneurial governance. Understanding these mandates, the local state initiated various experiments to explore new approaches to redevelopment. Some projects led by the local state absorb the principles into project operation. Nevertheless, the implementation of these government-funded micro-regeneration projects resorts to entrepreneurial means with local governance innovation and experimentalism. These general principles are adopted, and even utilized, by the local state to suit their practical needs, which may include making economic benefits from redevelopment projects.

Second, the current micro-redevelopment includes various local practices related to multiple national political mandates, including rural vitalization, the protection of rural landscapes, heritage conservation, ecological civilization, public participation, and the development of a harmonious society. Some projects may be initiated under a specific mandate. However, all incremental regeneration projects avoid large-scale demolition and dispossession, reflecting political intention. Although actual redevelopment practices may vary significantly, ranging from partial refurbishment, on-site restoration, renovation of community gardens and public spaces, creation of leisure and parking facilities to soft governance innovation for participation, they deliberately avoided residential relocation. Even when relocation is needed, compensation and resettlement are provided. These projects are not justified by profitability and revenue but by social outcomes.

Although we may argue that local officials use these mandates and discourses for their benefits and local growth,[13] these development motivations do not contradict political considerations, as the state mandate reflects an overall intention to maintain structural coherence. The territorial logic is not an independent force but rather subject to the political economy of capital accumulation. The government may even leverage its own resources to achieve developmentalism. Development capital may be drawn from fiscal expansion and various financial operations to fund redevelopment projects for state mandates. This process might not be just the 'financialization of urban redevelopment' (Rutland 2010).[14]

Third, these projects are operated through public–private partnerships. The developers take on actual construction work, while funding sources come from state-owned enterprises, state agencies, or partnerships. In the case of Yongqingfang, the district government solicited Vanke, a private sector development, to carry out the redevelopment project through BOT and contribute redevelopment capital. Although Vanke operates the project, the developer cannot capture land value appreciation. The private developer was motivated to expand its operations from real estate development to managing state projects.

Some micro-regeneration projects of dilapidated estates may not involve real estate developers. For example, Huajin in Wuhan was initiated by the district government of Wuchang but implemented through its street-level agency and the community. The real estate developer plays no role in decision-making. If there is a partnership, the state accommodates the reasonable interest of private developers. Although they have some discretion in developing deals, they are not dominant in the redevelopment project. In other words, these redevelopment projects are not led by developers. Unlike 'property-led redevelopment,' these projects may be achieved *through* property development.

Fourth, national political mandates and campaigns, alongside state politics, change over time in response to the specific problem and crisis at the historical moments. However, the overall trend is increasing state intervention in urban redevelopment. State politics shifted from promoting economic growth to maintaining 'common prosperity.' The changing mandates may affect and determine the viability of redevelopment projects.

Urban redevelopment projects may be initiated in response to specific policies. For example, Yongqingfang took the chance in response to a campaign promoting 'mass innovation and entrepreneurs.' Huajin in Wuhan gained access to government funding through the 'happy community' campaign. The campaign reflects an overall change in the discourse of urban redevelopment. One way to make local communities happier is to improve their physical living conditions. The central government sets the rationale for redevelopment, while local governments mobilize resources to respond to these initiatives.

Fifth, the national political mandates do not specify a concrete approach to project implementation. Therefore, these projects are not implemented in a top-down manner. Providing overall guidance, state politics permit local experiments, discretion, and innovation. Because the political mandates aim to achieve extra-economic objectives through economic means, they necessarily contain contradictions and dilemmas.

To fulfill these mandates, the local government has to breach existing institutional constraints, leading to greater informality. In urban redevelopment projects, there are various temporary arrangements. For example, farmers' land was borrowed for green space in the greenway development in the Pearl River Delta, and the local government allowed Liede in Guangzhou to redevelop their land and retain the 'collective land ownership' for building a five-star hotel. Yongqingfang in Guangzhou explores a new way of urban redevelopment. The district government used public housing areas to create 'mass innovation' spaces. This differs from Xintiandi in Shanghai, where the local government sold the site and nearby area to the developer—Shui On. In Yongqingfang, the government persuaded Vanke to conduct a redevelopment project for the government. The project is thus a 'creative destruction'—destroying the existing to create a new space for economic growth. It converted public housing to co-working offices rather than preserving the neighborhood. The developer receives the rent from these properties, but its profit from land is curtailed.

Nevertheless, the project of Yongqingfang attempts to exemplify itself as a 'cultural conservation district' through heritage planning and preservation. The creation of co-working spaces and a culture conservation district are all under their political mandates, which are not merely rhetoric. However, in a post-reform political economy, the project is an experiment that utilizes market instruments and mechanisms of public–private partnership. Even though the state's intention is not for real estate profitability and its role is forceful in urban redevelopment, the actual implementation led to an opposite outcome.

To sum up, although micro-redevelopment projects are beyond the local growth machine dynamics and initiated under the politics of multi-scalar state and national political mandates, they have significantly changed the neighborhoods. Vernacular rural villages are transformed into holiday resorts; traditional neighborhoods are converted into non-residential uses; dilapidated estates are upgraded but not regenerated through a self-mobilized society. These significant changes cannot be initiated by real estate projects alone. They are political

Figure 5.9 The development of Moshikou Historical and Cultural Streets in western Beijing aims to restore the historical trade route combined with new tourism, leisure, and multi-purpose residential complexes.

projects beyond the dynamics of real estate projects. Micro-regeneration combines economic and social changes through broader political strategies and state politics. Many Chinese cities widely see this combination of historical preservation, cultural branding, and commercial development (Figure 5.9).

Critical reflection

On gentrification

The concept of gentrification has been significantly expanded in the literature (Lees et al. 2016). Initially, it refers to the phenomenon in inner urban areas where the middle class moved into workers' neighborhoods. Wealthier middle-class newcomers replaced the working-class residents. Later, the concept was expanded to emphasize displacement; in fact, any displacement by residential or non-residential development, market, or non-market forms is the outcome of gentrification. For example, the change from residential to commercial uses can be called commercial gentrification. In the literature, gentrification is almost equivalent to displacement, or the two words are exchangeable—the nature of gentrification is displacement. However, a stricter view of gentrification does not regard all displacement as resulting from gentrification (Hamnett 2021).

In this sense, the example of Yongqingfang, a small-scale redevelopment in Guangzhou, can be classified as gentrification because, before the small-scale redevelopment, the site had already been vacated through residential relocation. The relocation resulted from real estate development. Displacement did occur. However, not all redevelopment led to displacement. Neighborhood refurbishment and rehabilitation in the old estate of Huajin in Wuhan did not see residential relocation and, hence, is not easily referred to as gentrification.

Does this mean that in China, the notion of gentrification exists in some places or cases but not in others, or that it is no longer prevalent? To fit these redevelopment practices into gentrification, we would need to further relax the definition to the extent that it is equivalent to any upgrading activities. Then, can all these cases of physical improvement by redevelopment be called gentrification? However, calling them gentrification does not enhance the understanding of the nature of these activities (Ren 2015; Ghertner 2014). Despite different concrete forms, these projects are all micro-redevelopment and reflect common governance features and rationales at particular historical moments, promoting economic growth, rehabilitating the living environment, and generating land revenue.

The redevelopment policy in China always intends to improve the living conditions of original residents. However, the regeneration of the larger area around Yongqingfang failed to achieve this purpose. It created the opposite outcome because few residents lived there during the shifting development approach. The project is not an aesthetic and lifestyle change by the middle class, but more for the state's political achievement for a better and preserved heritage environment according to its perception. The project is difficult to replicate, i.e., using property redevelopment to achieve its policy objectives. Financially, Yongqingfang would not have been viable if the state had not relocated these public housing tenants earlier. The financial contribution is still an issue for the housing estates that need refurbishment, like those in Wuhan, without changing the land uses. Despite the lack of intention to relocate them, micro-redevelopment also faces dilemmas. For property-led redevelopment, the challenge is social contestation and inequality. Although some micro-redevelopment projects pay attention to social issues, financing through property development becomes challenging, especially in the market downturn.

What we have seen in these cases is that they share some similarities with gentrification in market economies whereby the state considered or even pursued capital accumulation. Still, a territorial logic emerges as capital accumulation runs dialectically into its own contradiction—it is not a viable approach in the market term (Jessop 2002a). As a territorial logic to maintain structural coherence, the state experiments with its governance and shifts its emphasis along with the changes in historical conditions.

Gentrification as a concept powerfully reveals and stresses the first part of dynamics—the growth machine. Chinese urban redevelopment processes reveal the changing balance and state politics characterized by state entrepreneurialism. Thus, gentrification is still a useful concept as it illuminates an integral part of the dynamics of redevelopment in China. China's context of underdeveloped capitalism or yet-to-become capitalism helps to provide feedback on the other part of the dynamics, which is not at all absent in capitalism. For example, a recent discussion on municipal statecraft highlights urban development's complex state objectives and motivations (Lauermann 2018).

On neoliberalism

Is urban redevelopment the frontier of neoliberalism? Earlier urban redevelopment in China has been conceptualized as the frontier of neoliberalism (He and Wu, 2009). The introduction of market mechanisms into urban redevelopment in the 1990s has transformed the dynamics of redevelopment with a greater interest in property values. However, the role of the state is still significant (Wu and Zhang, 2022). Small-scale redevelopment and the regeneration of dilapidated areas reveal some new practices. It would be theoretically challenging to ask whether these practices still present neoliberalism. Some may still show such a feature as the state outsources the redevelopment task to the market and encourages the residents to be responsible for themselves—creating a new subjectivity through individualism.

When we closely examine the notion of neoliberalism in China, we find that neoliberalism has been a survival strategy for the state in response to internal and external challenges (Wu 2010). The market has been introduced as a practical solution to the crisis of capital accumulation—difficult to maintain economic growth in a state-owned planned economy. In urban redevelopment, property-led redevelopment became dominant along with the property boom in the 1990s and 2000s. The state has initiated this change, or, more precisely, through an exception of its regulation as a particular form of governmentality (Ong 2006). It presents a perplexing market rationale—neoliberalism—in urban redevelopment. However, behind the market operation is always state strategy.

Nevertheless, the state has to consider the capital logic by using the approach of 'property-led redevelopment.' Such a capital logic has led to new contradictions, social contestation, and social and financial crises. To maintain an overall structural coherence, the state has introduced new redevelopment practices—micro-regeneration— which reinstalls social objectives, for example, to improve the living conditions of the poor, to create jobs, preserve heritage, and make the people happier, even though these discourses may not achieve their purposes. Some major redevelopment projects are associated with a wider consideration of economic restructuring and urban development (Figure 5.10). Therefore, while redevelopment has shifted from property-led redevelopment to micro-redevelopment, state entrepreneurialism has been consistent through these phases. The same logic presents different policy emphases according to the perceived urgency at the specific historical moments.

On property-led redevelopment and state entrepreneurialism

In sum, this chapter focuses on concrete urban redevelopment approaches, through which we try to understand urban governance. According to Harvey (1989), urban governance has transformed from managerialism to entrepreneurialism in Western market economies. Such a shift is still within the capital logic of capitalism, reflecting different concrete forms of governance. This shift has changed urban redevelopment practices from urban renewal to property-led redevelopment and variegated approaches to regeneration. Although the concept of property-led redevelopment is quite ambiguous, the concept at least indicates the use of property development—the development of office buildings and commercial uses in regenerating old and derelict areas. Property developers play a more significant role in this type of redevelopment, alone or through public–private partnerships. This approach leads to gentrification and displacement.

China is not an exception. Similarly, it introduced property-led redevelopment, by which urban redevelopment has become the frontier of market experiments—neoliberalism. However, the state still plays a significant role in demolition and facilitating urban redevelopment, as shown in urban village redevelopment. Besides the rationale of profit-making, it demonstrates its intention to maintain social order and remove informalities caused by villagers' self-built activities.

Urban villages are quite strong local communities because of the history of the lineage. Through the shareholding reform of village land, the village actually manages its collective assets through a quasi-market form. In other words, the state permits local communities to

Figure 5.10 The redevelopment of Shougang Park, on the site of iron and steel production in Beijing, preserves and utilizes its industrial heritage.

deploy a market approach. The state sometimes compromises its interests in land value capture (Lin 2015)—as the value capture is often used as a means to regenerate the built environment. If the villages are weaker, they usually face the fate of demolition. The state deploys the market form for them. Legally developed spaces are compensated in land acquisition.

However, in some cases, even illegal spaces received compensation under a pragmatic attitude to achieve the larger objective of the state. The state may even allow villagers to develop a higher density to achieve its village redevelopment purpose. The practice of old city and village redevelopment demonstrates such a dynamic, indicating a new phase of redevelopment that gives the local community greater discretion to deploy market approaches. This is different from the earlier and common practice of the state deploying the developer to regenerate while providing a more outstanding preferential treatment to developers' profit-making motivation.

In this particular historical moment, the old village redevelopment gives more incentives directly to local communities. This new phase further reveals the nature of state entrepreneurialism, which has been less visible because of the practical consideration of market operation. More recently, large-scale demolition has been discouraged. Small-scale development or micro-regeneration has been experimented with developing creative industries and heritage conservation in traditional urban neighborhoods. In dilapidated housing estates, partial regeneration and refurbishment are encouraged. We can learn from these new practices that, unlike the earlier practices, they demonstrate a visible aspect of state entrepreneurialism. Property-led redevelopment and gentrification, in their strict definition in the Western context of urban studies, might not capture the full dynamics. These concepts have to be significantly modified. Chinese practices encourage us to rethink their meanings.

Notes

1 For a comprehensive introduction to urban redevelopment and regeneration practices in the UK, please see Roberts et al. (2016), Tallon (2021).
2 See Hamnett and Shoval (2003) for the role of museums as flagships of urban development.
3 There is an extensive literature on the case of Xingtiandi, see He and Wu (2005), Yang and Chang (2007), Ren (2008), Wu (2022a).
4 This section is developed from Wu (2020b).
5 The book provides a wider context of neighborhood changes as the production of different spatial governance forms. See also Wu (2020b) for the use of gentrification in Chinese and Indian cities.
6 See for villagers' property development activities in North-eastern China.
7 This new central business district is the Pearl River New Town.
8 See also Liu et al. (2023) for the policy of *sanjiu* redevelopment in Guangdong.
9 See Gilbert (2007) for a critique of the language of slum to describe these informal settlements. See Myers (2020) for comparative studies on informal development and urbanism in the Global South.
10 The case has been extensively studied, see also Wang et al. (2022a, 2022b), Yu et al. (2021), He and Qian (2023).
11 See also Li (2022) for another case in Beijing.
12 This section is developed from Wu et al. (2022b).
13 These can be seen in many environmental projects, see Chung and Xu (2021).
14 Therefore, it is essential to critically unpack the meaning of financialization (Christophers 2015). In the context of fiscal austerity, financialization represents an approach to extracting value from urban redevelopment. Financial instruments are deployed to fund rehabilitation projects in other contexts where the political imperative becomes more apparent. Housing development in China can also mobilize capital for economic development and public finance (see Wu et al. 2020).

6 Innovation

A hybrid 'national indigenous' model

Introduction

Between China's access to the World Trade Organization (WTO) in 2001 and the global financial crisis in 2008, its economy achieved remarkable growth. Through the introduction of market mechanisms to its urbanization and science and technology (S&T) development since the 1980s, China has attracted global capital to its manufacturing industries and multinational R&D centers. However, its use of the domestic market in exchange for technology has proven less successful (Zhou et al. 2016).[1] The Chinese economy in the mid-2000s was stuck in labor-intensive, low-value-added manufacturing industries. Although the state makes constant efforts and targeted programs through 'experimental governance' to foster innovation (Heilmann et al. 2013), the S&T system still faces significant constraints. Because of its economic structure, domestic enterprises have avoided investment in innovative and risky technologies. In the mid-2000s, labor cost increases began to impact labor-intensive manufacturing.

Thereafter, the global financial crisis seriously threatened the export-oriented economy. Thus, promoting economic restructuring and moving up the value chain became a priority in governing economic development. China has strived to increase R&D investment, raising the ratio of R&D spending to GDP from 1.32 percent in 2005 to 2.5 percent in 2020 (Fu et al. 2021).

China is now the second-largest economy in the world, with R&D spending second only to the US (Zhou et al. 2016; Appelbaum et al. 2018; Fu et al. 2021). Some large corporations, such as Huawei, Tencent, Alibaba, and Xiaomi, are now renowned around the world. China makes persistent use of industrial policies to support key industries. Its institutional capacity is a legacy of the planned economy. However, recently, the guidance of the market and industrialization resembles the experience of the developmental state in East Asia. To a lesser extent, governing economic development is understood through urban and spatial policies in addition to industrial policy (Zhu et al. 2023; Zhu et al. 2024). These policies to govern innovation identify the key strategic industrial sector and designate industrial development zones and high-tech parks—they are similar to the so-called 'cluster policies'—strategies and initiatives to foster innovation through firm agglomerations (Porter 2000).

We must investigate national and regional innovation systems in order to understand their respective innovation models. Such models are closely related to economic development and are subject to geopolitical conditions. The development may focus on export-oriented manufacturing or high-quality and innovation-driven development in different historical periods. These policy choices of state entrepreneurialism aim to overcome development constraints in the historical moments. This chapter argues that China's innovation governance model is a hybrid under the overall policy of state entrepreneurialism, conceptualized in Chapter 1 on

DOI: 10.4324/9781003355175-7

governance. With the post-pandemic era of intensified US–China conflict and the resulting changes in global geopolitics, China seems to strengthen the state's strategic guidance over innovation. We will detail how this model operates to govern innovation spaces.

National innovation models

Three distinctive national innovation models are developed associated with 'varieties of capitalism' (VoC) (Hall and Soskice 2001;, Cooke et al. 2007, pp. 12–131; Zhang et al. 2011, pp. 577–579). First, the entrepreneurial model is under the liberal market, with market mechanisms coordinating the economy. Corporate governance in this model is based on private ownership, allowing high flexibility and enabling a swift response to market signals. A competitive ethos, individualism, and market rationality drive innovation. This model emphasizes creative or scientific novelty hedged by intellectual property rights. The flexible labor market allows job mobility. The local economy, especially urban creative and high-tech industries, strives to attract talent. Abundant opportunities in the region create an agglomeration effect in that location. The R&D system is based on universities and research institutes. Although the state might sponsor defense and healthcare research, their implementation is based on market selection. The state does not substitute for market-initiated R&D and venture capital plays a more critical role in financing innovation. The state programs are executed through the market and R&D is based on the private sector. The development of high-tech industries uses the business strategy of outsourcing and subcontracting based on competitive tendering, thereby maximizing efficiency. The US model is considered a typically entrepreneurial one, whereas the UK is a variant on it.

Second, the associative or public–private partnership model is based on rules and the legalistic regulation of economic activity. While the market mechanism is underlying innovation activities, it is subject to a visible regulatory state. In this model the state negotiates with private and industry associations and delegates economic functions to them. The state also requires private entities to take certain social functions. The business and industry associations are key intermediaries facilitating collaborations between competitive private entities. Business partners are established through the associations. The state regulatory function also devolves into self-management in local economies. The local political environment is a stable one. Negotiation can also occur across jurisdictional boundaries. Government, industry, and labor have divergent interests. The partnership model reinforces long-term cooperation among these economic actors, leading to the continuous exploitation of local partnerships and interfirm linkages. Companies share the risks of innovation. The associative model permits collaboration in addition to competition. The R&D system involves universities and industries in cooperation. Industry associations and other intermediaries, such as chambers of commerce and industry, assist with training and technical support. The investment in innovation is more credit-based, as is the overall capital market. With close links between banks and companies, financial institutions provide long-term investment and act as patient capital. Cooperation between large and smaller enterprises enables innovation gains from research findings to filter through the innovation chain rather than awaiting mainly entrepreneurial impulse from market signals because the business association coordinates innovation activities. The government plays an important role in developing region-based industrial clusters through innovation policies. This partnership model is used in many continental European economies and also in the Nordic countries.

Third, the developmental state model, which differs from the liberal and coordinated markets, is conceived from the system of East Asian state-centric economic coordination. In this, the state follows a plan rationale rather than a market rationale. A strong and authoritarian central government characterizes the model. State strategy can be implemented through a multi-scalar state that maintains its authority through top-down appointments. The government strategically supports large enterprises in order to strengthen national industrial competitiveness. Large corporations play an important role in innovation in the developmental state model. Within this model, there are also a number of variations. For example, Japan is more a case of state-guided coordination, as the state encourages the banking sector to support strategic industrial sectors.

By contrast, South Korea is a 'state-organized' system, as the state involves large corporations in the private sector. China is also considered a variant of the developmental state because it achieves its development goals through state-owned enterprises. In China, the state controls ownership through state-owned enterprises and the capital market, as banks are primarily state-owned, and provide favorable financial support to targeted and key enterprises.

The linkage between university and industry is weaker in the developmental state model than in the associative model. The latter system often sees more interactions between government, universities, and industries (something known as the 'triple helix' of innovation) (Etzkowitz and Leydesdorff 2000). The innovation system lacks spin-offs and technology-oriented start-ups in the developmental state model. Because of the limited public funding for basic research, and the relatively weak position of universities in the national innovation system, science-based industries within this model are relatively weak compared to those in the entrepreneurial model. It has a significant constraint on science-intensive innovation such as biotechnology.

These three models all have their own respective strengths and weaknesses in promoting innovation. The entrepreneurial model is more flexible, can respond to market signals, and adjusts quickly and effectively. The partnership model maintains a long-term strategy and may outperform the liberal market during a stable economic and global trading environment. The developmental state model can achieve even longer strategic objectives and develop new industries and key sectors for industrialization beyond path dependence. Through a top-down policy, the model implements a long-term national development vision.

Beyond the strengths and weaknesses of their approaches to innovation, there is also some criticism of the methodology of understanding innovation using these models. As varieties of capitalism, these models are ideal types and each can be criticized for their 'methodological nationalism' because they describe distinctive approaches based on their nation-state without considering hybrid and combinational approaches. For example, in a commonly noted entrepreneurial governance, the 'entrepreneurial state' plays a significant role in defense innovation through a 'mission-oriented' approach (Mazzucato 2013; O'Mara 2020; Zukin 2020). These models are also categorized rather than theorized. However, there are various mixed and combinational approaches; hence, creating these categories might not help in understanding the underlying mechanism of innovation.

How does China fit into these three models? It is tempting to associate China with the developmental state model because of the decisive role in development played by the state. However, as will be seen later through detailed and fine-grained studies of innovation spaces in Chinese cities, the nation's innovation approach does not easily fit into any of these models. The Chinese approach contains all elements of these models: enterprise-centered innovation with market development; partnerships between local states and development corporations and lead firms; and multi-scalar states that fulfill national innovation strategies.

Recent studies of China's innovation system attempt to understand the reinvented role of the Chinese state in promoting innovation within the literature of 'state capitalism,' creating the 'new whole state system' as an innovation model (Zhang and Lan 2023).[2] These studies do not attempt to contrast 'Western democratic, free-market capitalism' with 'Eastern authoritarian state capitalism.' From the global production networks (GPNs) studies in East Asia, the role of the lead firm in GPNs is stressed, as the developmental state itself is no longer a driving force. They suggest that neither neoliberalism nor the developmental state is the source of innovation (Yeung 2019b). Therefore, these models are not regarded as a good theory for innovation, and there are various attempts to theorize innovation, as explained below.

Concepts in economic geography

The prevailing paradigms, which explain the emergence of innovations, can be broadly divided into territorial (localities) and relational ones. The territorial one primarily focuses on the local or regional characteristics and assets, while the relational one pays attention to interconnected places, primarily through production networks. While some theories bridge the local and relational understanding of innovation, such as the theory of 'local buzz and global pipelines' (Bathelt et al. 2004) and 'strategic coupling' (MacKinnon 2012; Coe and Yeung 2019), with attention to local features, they have in essence a network view. In addition to a more static view of regional assets, there is an 'evolutionary economic geography' (Boschma and Frenken 2006). The relational paradigm is developed by critiquing territorial or local ones—the 'cluster theory' (Porter 2000). The evolutionary view can include the critiques of fixed regional assets and geographical proximity (Boschma 2005) and a dynamic view of GPNs (MacKinnon 2012). The territorial and relation paradigms are extended into respective dynamic and evolutionary views.

These two paradigmatic explanations of innovation focus on knowledge creation in innovation, which does not regard governance change as part of innovation. Although they believe that innovations are situated in the context of governance—different innovation models as discussed earlier in this chapter, they do not focus on the actors—entrepreneurs, the business association, or the government, with the exception that the theory of GPNs tends to focus on the lead firm. To a lesser extent, they reveal the agency.[3] They are not oriented toward explaining the role of the state, market, or society in innovation. In essence, it seems that existing economic geography tends to treat the political economy as the context for innovation rather than a driver intrinsically associated with or part of the dynamics of geopolitics, with a few exceptions, especially under state capitalism (Glassman and Choi 2014).

This book examines the governance of innovation rather than innovation itself. In other words, we look at governing innovation activities alongside innovation economies. It is a political economy rather than a knowledge model. We take a brief step back to explore the process outside knowledge production, including urban development, planning, and state and spatial politics. They constitute broader innovation processes. Here, we adopt a political economy view of innovation processes.

The existing paradigm of local interaction seeks to identify the factors of a place that are believed to be conducive to knowledge generation. Agglomeration is recognized as a significant factor by the cluster theory, which emphasizes the self-reinforced effect of geographical concentration—conceived as 'spatial proximity' (Boschma 2005). More recently, the city's contribution to innovation is understood in term of its agglomeration nature (Scott and Storper 2015). At a regional level, the understanding of agglomeration turns into a 'new regionalism.' However, such a description of industrial districts is criticized as policy advocacy without a sound understanding of its dynamics (Lovering 1999). The cluster theory is criticized for its excessive attention to spatial proximity, especially the closeness of the sameness of firms.

Also, focusing on the locals, a different explanation emphasizes diversity. At the industrial sector and firm levels, the theory of 'related variety' suggests that the heterogeneity rather than the homogeneity of cluster functions facilitates the creation of new knowledge (Boschma et al. 2014). Diversity is regarded as an essential determinant of innovation at the city level, where innovations are located. This is explained by the chance of unexpected encounters and face-to-face contacts associated with 'buzz,' which enhances innovation (Storper and Venables 2004). According to Schumpeter's theory, the city is seen as an 'innovation machine' because it combines the city's diversity, as Jane Jacobs (1970) argued, and entrepreneurship (Florida et al. 2017). Related diversity focuses on the knowledge dynamics generated by diverse but related economic sectors or firms. Implicitly, like the cluster theory, it provides a different policy recommendation, i.e., to co-locate heterogeneous firms that can facilitate innovation. To a lesser extent, it asks the question of how diverse firms in various sectors come to co-locate in the same locality in the first place—whether this is out of market dynamics, social and cultural behavior, or conscious actions by the state innovation policy, or a mix of some or all these actors.

The regional innovation system (RIS) concept emphasizes a broader range of environmental attributes in addition to the spatial proximity of firms (Cooke 1992). These attributes include institutions and local culture (Cooke and Morgan 1998). RIS includes not only geographies but also regional assets—or conditions. For example, social capital and trust are deemed important factors of RIS. The interactive relationship between university, industry, and government is conceived as the 'triple helix model' for innovation (Etzkowitz and Leydesdorff 2000). The extension from clustering to the RIS model enriches the scope of the investigation. It opens up the possibility of interrogating the urban development process and furthering the political economy in which innovation occurs. However, both clustering and RIS (or the triple helix model) essentially follow a territorial paradigm, confining their scope to localities. All these explanations pay attention to local or regional assets. The relational turn of economic geography pays more attention to extra-regional connections and an external environment outside clusters or a broader context of innovation and its governance.

An extension of RIS is a prescriptive 'platform policy' (Asheim et al. 2011), conceived by three key scholars in RIS and related variety.[4] The platform policy still focuses on knowledge bases and dynamics. The policy promotes constructing a platform—presumably at the exact location or localities—of differentiated and diverse knowledge bases to achieve 'related variety' using the co-location of various sectors to cross-fertilize knowledge. The policy is more sophisticated than clustering, but is in the same spirit of knowledge interaction. The platform distinguishes knowledge into 'analytical science-based,' 'synthetic and engineering based,' and 'symbolic and arts-based' in nature to build different prolixity mixes to suit their needs (Asheim et al. 2011).

Moving from the territorial view of innovation to a relational view, greater attention has been paid to the network. The 'local buzz and global pipelines' model combines local interactions and trans-local linkages (Bathelt et al. 2004). The 'local buzz' refers to local information and communication ecology, while 'global pipelines' connect local innovations to other places. The pipeline mainly refers to knowledge flows and dynamics. The theory focuses on knowledge connections between these places, which differs from a wider 'conjuncture' view that relates different places in their historical moments (Peck 2015). In this sense, the 'local-buzz and global pipelines' model is a defined economic geography model of innovation rather than a political economy theory of capitalism that survives and is centered upon innovation. The latter concerns the fundamental feature of so-called uneven development.

Beyond the knowledge dynamics, the GPN perspective extends innovation in the context of trans-local connections in production, explaining how these production activities in different localities are related across the world at a global scale. Thus, GPN relates globalization and regional changes. While the theories of local buzz, global pipeline, and GPN pay attention to the network, they each have a different focus. The former pays attention to the knowledge network, while the latter adopts the scope of broader production activities.

GPN's perspective tends to focus on the network. Recent development extends its attention to the localities through the concept of 'strategic coupling' (MacKinnon 2012), which investigates how global lead firms in GPN are related to regional assets or how they strategically select localities with assets that suit GPN's needs. In theory, the process is two-way. The localities can strategically select their desired GPNs by embedding the local economy into GPNs through the industrial chain chief in China (Gong et al. 2022).

Further development is needed from a dynamic point of view of GPN. The mismatch between GPN and local assets may lead to 'decoupling,' and additional changes may lead to 're-coupling' (MacKinnon 2012). These concepts are intuitive, describing the changing relation between GPN and localities. The concepts are also flexible, leaving to the explanation of the forces driving the changes in GPNs and localities and their relations to empirical observations. The GPN and strategic coupling perspectives have been applied to China (Yang 2014a, 2014b; Fu and Lim 2022; Gong et al. 2022).

To sum up, although these theories provide contrasting explanations from agglomeration to diversity, they are primarily concerned with the knowledge dynamics of innovation in localities. The theories of local buzz and global pipelines and strategic coupling of GPN go beyond the locality and pay attention to global and local interactions. These theories do not preclude policy prescription, and their policy implication might be straightforward from their respective theoretical positions. They examine the dynamic innovation process as interactions between regional assets and GPN operations. These are closely related to an innovation policy or lead firms' decisions. Their scope is more on knowledge development and less on urban development within which innovation spaces are constructed and knowledge dynamics occur. Situating innovation governance in these historical and material city-building processes is helpful.

To understand these processes, we need to investigate urban and regional development. While innovations primarily occur in specific localities like the city, the city is not an actor or has an agency. It is where many actors play their roles related to the knowledge dynamics directly or indirectly, such as financing, land ownership, and public policy formation, which contribute to governing innovation. These are not just innovation or industrial policies, but also a wide range of public policies and geopolitics considerations. These actors may be outside the locality—rightly perceived by the perspective of GPN, and act in a trans-scalar way, but trans-scalar actors are not limited to the lead firms of GPNs (Halbert and Rouanet 2014).

From this trans-scalar perspective, the city is an assemblage of various actors working together in this local space. Their trans-scalar activities and influences are broadly situated in the political economy. Understanding innovation governance through the political economy view, Silicon Valley thus represents a particular development of knowledge economies related to generous funding from US federal military contracts (Mazzucato 2013; O'Mara 2020; Zukin 2020). Another example is the development of South Korean ICT out of Cold War geopolitics (Glassman and Choi 2014). The city may not be the only spatial form that contains or generates innovation (Shearmur 2012). In many circumstances, it is simply a 'key geographical locus' produced by many other agents, leading to the development of innovations in cities (Shearmur 2012). In China, the political economy of urbanization and rescaling through decentralization led to the city on a substantial scale where major development is concentrated (Wu and Zhang 2022). We now introduce the historical innovation development process in the cities and their high-tech parks as concrete outcomes of the development process.

Changing innovation models in China

China's innovation system has experienced a significant transition since the economic reform. After introducing the market mechanism, its innovation model changed from being mission-oriented, funded by the central government and targeted at key industrial sectors, to more evolved local innovation systems based on industrial enterprises (Liu and White 2001; Zhang and Wu 2012). Globalization and China's opening to the world attracted foreign direct investment and the development of global R&D centers in its major cities such as Shanghai.

Despite the evolution of its innovation model, and the transition to one with a response to the market signal, China has not converged into a liberal entrepreneurial model. While industries become active actors in innovation, industrial associations or local institutions do not lead innovation because China entered the global value chain through labor-intensive manufacturing, showing path dependence. China differs from the associative or partnership model. China resembles the developmental state model, in which state industrial policies provide a strategic direction for innovation.

Nevertheless, China also has a much-devolved multi-scalar state, as discussed in Chapter 1. For example, high-tech development zones have been established through experimental governance under fragmented authoritarianism (Heilmann et al. 2013). In the transition period from 1978 to 2008, when the capital accumulation regime of the World Workshop operated, its regulation mode of innovation was remarkably market-friendly. However, since the crisis of financial capitalism and the intensified geopolitical tension, or what might even be termed a 'new Cold War,' innovation again arose as an issue of national security. The strategic role of the state has been revived. In this respect, China's increasingly visible state has been shaped by the political economy of development and geopolitics, sharing some features with the East Asian developmental state as the governance outcome of the Cold War (Glassman and Choi 2014, Yeung 2017). They have been driven by geopolitics and financialized global capitalism.

From the 1950s to the 1970s: mission-oriented innovation with dominant government funding

The socialist and pre-reform innovation governance was characterized by a highly centralized and hierarchical structure controlled by complete state ownership of R&D (Liu and White 2001). In 1956, at the height of the Cold War, the National Science and Technology Long-term Plan, which focused on the development of atomic energy, electronics, semiconductors, automation, and rocket technology, was formulated through a top-down approach. The state was the driver of innovation, centered on the mission to develop defense industries and technologies. To cope with the constraints of its weak economic capacity, the state quickly assembled the resources of R&D to accomplish a critical mission. This led to compartmentalization, which hindered cross-fertilization and innovation (Liu and White 2001).

From the 1980s to the mid-2000s: enterprise-centered innovation under transition to the market economy

With the rise of the US-led neoliberal international political environment, and in particular China entering the WTO, the country shifted toward economic growth. It followed the East Asian development experience and absorbed foreign direct investment through labor-intensive industries. China encouraged collaboration between foreign investment and also the use of foreign technologies. It hoped to use foreign investment in a higher technological area by offering access to the domestic market. Multinational corporations set up their R&D centers. China hoped to import foreign technology through the 'open door' to its domestic market. However, using the domestic market to gain technology has largely failed (Zhou 2008a). Foreign investment dominates high-tech exports. The progress of technological innovation under the impact of foreign direct investment has been slower than expected. Foreign firms monopolized China's export market, and domestic firms lagged behind.

The state promoted commercialization and encouraged the formation of inter-city knowledge linkages. However, innovation is still highly uneven. Because of local knowledge assets, global R&D centers are concentrated in large cities such as Beijing and Shanghai. In this period, China also saw the emergence of ICT clusters under flexible entrepreneurial urban governance and strengthened university–firm links in the form of university-affiliated enterprises (Wu 2010; Lyu et al. 2019). Chinese cities participated in the GPN. For example, various state authorities, from the central to municipal governments, made a collective effort to encourage explicit coupling with global lead firms; this can be seen in Shenzhen's new liquid crystal display industry (Yang 2014b). In this period, China set up a number of high-tech parks.

From the mid-2000s to present: state strategic guidance over innovation under state entrepreneurialism

The Medium- and Long-Term Plan for the Development of Science and Technology, introduced in 2006, initiated 16 mega-projects to enhance indigenous innovation capacities. The gradual rise of labor costs from the mid-2000s has already propelled some labor-intensive industries to relocate from the coastal region to inland and other Southeast Asian countries.

For a long time, China's technological innovation has been constrained by the lack of capital. However, as discussed in Chapter 3, recent crisis-triggered financialization means that China can use debt finance to invest in R&D. The global financial crisis in 2008 significantly impacted China's export-oriented industries. The central government launched a financial stimulus package of 4 trillion Yuan to stabilize the economy. The crisis exposed the vulnerability of low-value-added manufacturing industries.

The Chinese government aimed to promote economic restructuring toward developing high-value-added industries. Faced with the constraint of external markets for labor-intensive industries, the stimulus package invests in infrastructure development. It did not immediately require an expanded consumer market, and 10 percent of the budget was allocated to innovation. As mentioned in Chapter 3, the stimulus package was achieved through local financial platforms—through financial means. In other words, the multi-scalar state internally generated and mobilized the capital. This has further strengthened the role of the state in economic governance.

The central government identified nine industrial sectors, including equipment manufacturing and ICT with high-tech components, as the primary investment areas. Investment under the stimulus package focused on the development of 'high and new technological zones.' With a defined boundary, these zones were designed to receive the investment under the stimulus package. The strategy was not about governing innovation per se, but rather about using state investment to promote economic development, absorb surplus labor, and stabilize the economy following the global financial crisis. More precisely, the zones are intended to raise the development capital—using land as assets to finance economic development, as shown in Chapter 3. Since 2008, the number of high-tech zones has significantly increased, reaching 178 in 2023. The government has also strengthened intellectual property protection and begun to certify high-tech firms. By the end of 2023, some 23 indigenous innovation demonstration zones had been established to promote the upgrading of development zones. The first four are Zhongguancun in Beijing, Donghu in Wuhan, Zhangjiang in Shanghai, and Shenzhen.

In 2012, the central government identified the 'strategic emerging industries,' and in 2015, they also launched 'Made in China 2025' (Yeung 2019a), which aimed to develop China's high-tech manufacturing industries. China's strategic intervention through proactive industrial policies is a salient feature of its economic governance, within which innovation is coordinated. Governing innovation is part and parcel of economic governance.

Despite substantial financial injections and escaping from an economic crisis, the Chinese economy entered a period of slow growth—which has been called the 'new normal.' The infrastructure-led development and an approach of financialization to economic growth—development by borrowing through local government financial vehicles (Pan et al. 2017)—created two significant challenges in the new normal: redundant industrial capacities in the constrained global and domestic market and local government debt.

Underneath these problems is a new crisis of capital accumulation—interpreted by the Chinese government as a structural problem. The central government thus initiated the 'supply-side reform' to reduce the overcapacity in low-value-added manufacturing sectors and pursue 'high-quality development.' The new situation of economic restructuring further raised the imperative of innovation. In the 13th Five-Year Plan beginning in 2016, the central government stressed the policy of 'innovation-driven development,' meaning that innovation should be an overall economic development strategy and integrated into urbanization. Four 'national comprehensive innovation centers', including Shanghai (2016), Hefei and Beijing (2017), and Shenzhen (2020), were designated to implement the innovation-driven strategy. These centers are, in essence, a city-wide initiative, not limited to a development zone within these metropolises.

These cities receive crucial infrastructure investment from the central government, such as the large national research infrastructure and the critical state laboratories. Further, they transformed high-tech parks or created new 'science cities' to accommodate investment. For example, Zhangjiang High-tech Park in Shanghai has been transformed into Zhangjiang Science City in an attempt to implement the national innovation strategy. Under this development, the area has been extended from 94 square kilometers to 220 square kilometers, indicating that the science city is now a completely new city within Shanghai. The development is integrated with the Pudong New District. The Science City focuses on strategic and emerging industries, such as integrated circuits, biotech, and artificial intelligence (Zhang 2015, Zhu et al. 2023).

Since 2018, the vulnerability of the Chinese economy has been exposed under the intensified US–China trade conflict and geopolitical changes. The chip embargo raises the Chinese government's determination to pursue its innovation capacities. This external pressure revived the old mission-oriented model into a 'new whole state system' (Zhang and Lan 2023). In the selected core technologies and strategic industries, the role of the Chinese state has been reinvented in order to promote innovation. However, in contrast to the old model, which relied on state funding, the new model finances innovation through a financial approach involving private-sector collaboration. The new practices under this state-led model include the mixed-ownership reform of core innovation firms, government-guided funds (Pan et al. 2021), and local government financing vehicles—market actors who had previously been profit-seeking changed their behavior with consolidated party-state leadership in enterprises. The state hoped they would become the agents of state strategies, as shown in the enterprises affiliated with universities (Zhang and Lan 2023).

Summary: urban-based innovation as a strategy

China's innovation model is fluid, evolving in response to external pressure and opportunities and internal challenges and crises. It changed from an initial state-funded, mission-oriented approach to a greater reliance on the market mechanism and actors. Through various policy experiments such as the 'Torch Program' and high-tech development zones, the state pursued innovation, but, as a result of the lack of capital, its R&D investment was somewhat constrained. With China joining global production and introducing market mechanisms, its innovation has become increasingly enterprise-centered. While the industry was characterized by low value-added sectors with little investment in R&D, joining the WTO has led to the inflow of new capital to build global R&D centers in major Chinese cities.

However, foreign direct investment generated limited innovation spillover, partially because of China's low-end economic structure and partially because of the lack of core technologies in high-value-added industries. The investment in global R&D centers was mainly in science experiments, such as contract research organizations in biotech to serve multinationals (Zhang and Wu 2012). The innovation under foreign investment does not lead to upgrading the economic structure toward high-tech industries in China, as the country still lacks control over intellectual property rights. However, global R&D centers still provide some positive effects as they provide jobs to local technicians and scientists. They also attract overseas returnees, who may later transfer to other enterprises. Foreign investment also upgraded the export structure with more technological components.

Despite increased R&D to GDP, the Chinese economy has been severely constrained by the lack of innovation investment. China's highly trained labor force, especially its technicians and scientists, is still cheaper compared to more developed economies. This suggests China's potential to become a knowledge economy and play a more significant role in global innovation. The Chinese government recognized the structural problem, as the country's export-oriented economy was at the bottom of the value chain and hoped to move up. The rapid development in the 2000s and rising land prices and labor costs meant a rising imperative for economic upgrading. This led to gradually formulating an 'indigenous innovation' policy to gain a more significant share in intellectual property rights.

The global financial crisis in 2008 was a significant trigger for the shift of the innovation model because the previous development zone model was in crisis. Under the export-oriented development, even the high- and new technology zones were not sufficiently oriented toward high-tech. They engaged in expansive land and property development. More innovation-intensive high-tech parks are limited to a few, including Zhongguancun in Beijing, Zhangjiang in Shanghai, Optics Valley in Wuhan (Geng et al. 2023), and districts in Shenzhen. The 'indigenous innovation' policy does not exclude the participation of foreign investment firms. Instead, the policy encourages the development of high-tech parks, which hopes to help China move up the value chain through innovation. Since the Global Financial Crisis, China has adopted a financialization approach to raising its development capital. This shift is profound because it enables the state to participate more actively in innovation. Urban financialization extends state power in governing innovation, as seen in some strategic investments by the local government of Hefei in leading ICT firms.

The history of the evolution of China's innovation policy shows that the model responds to a series of crises and challenges. Although the state intervened to a different extent in these stages, it has consistently tried to upgrade the economy through proactive industrial policies to innovation-driven urban development. Zonal technology was used to govern innovation. Increasingly, however, these high-tech zones, or the Chinese version of the 'science city,' embody state strategic intention rather than an exception of state governance. For example,

Figure 6.1 The Urban Computing Center acts as the city brain in Xiong'an New Area.

an urban computing center, named the 'city brain,' has been built inside the national strategic new area of Xiong'an (Figure 6.1). The strategy combines the development of a new city and technological innovation. State involvement is a material condition on which capital circulation has been sustained. For example, the state strategically invests in emerging green industries and smart technologies, providing market conditions and stability to cushion external impacts (Zhang et al. 2021, Xu et al. 2022).

China now has a more hybrid approach, combining state strategies and market means. Despite the startling difference in appearance, China's innovation model is more similar to the US defense-initiated but market-implemented innovation approach (Mazzucato 2013, Zukin 2020). In this sense, China is not an exceptional model but reveals the evolving innovation model under state capitalism. The difference is perhaps a multi-scalar state in its strategy formulation and a more significant proportion of state-owned enterprises. Nevertheless, its innovation model is still enterprise-centered rather than state-funded and mission-oriented. In the next section, we further elaborate on this urban-based strategy.

China's innovation space: the high-tech park

One prominent feature of China's innovation is uneven development and concentration in designated high-tech parks, especially in first-tier cities. One of the exemplars is Zhongguancun in Beijing. It originated from the marketization of state resources and public research institutes in the early stage of economic reform. ITC firms clustered along Zhongguancun Street. They collaborated with multinationals, learned their experience, and adapted to the domestic market (Zhou and Tong 2003; Zhou 2008a). Another well-known high-tech development zone is Zhangjiang High-tech Park in Shanghai. In this section, we explain the development of Zhangjiang in detail.

From the high-tech park to the science city: an urban strategy of Zhangjiang

Zhangjiang High-Tech Park (ZJHP) in Shanghai was established in 1992. It is located in the Pudong New District, which was developed as one of the four development zones in Pudong. Between the inner and outer ring road, this peri-urban location is halfway from the city proper of Shanghai, and also a half-hour drive from Pudong International Airport. It has a good transport connection with Shanghai and a maglev to the airport. The high-tech park is primarily a suburban, low-density development zone built on former rural land. The development of Zhangjiang was slow in the 1990s. In 1999, Shanghai developed a 'Focusing on Zhangjiang' strategy, identifying biotechnology, integrated circuits, and software as key high-tech industries. In 2011, Zhangjiang was designated as one of three 'National Indigenous Innovation Demonstration Zones' following the publication of the Medium- to Long-Term Strategic Plan for the Development of Science and Technology (MLP) in 2006.

In 2014, President Xi Jinping visited Shanghai, hoping that the city would develop itself as a 'global science and technological innovation center.' To achieve this mission, the State Council established a leading group intended to guide national innovation and grant the establishment of four 'National Comprehensive Innovation Centers' in 2016. In the same year, Zhangjiang was named the science city as an operational space to establish Shanghai as a national comprehensive innovation center. As a result, the master plan of Zhangjiang Science City (2017–2035) was prepared. Zhangjiang has changed from a development zone (a park with high-tech industries) to a sub-center of Shanghai, and its jurisdiction area was expanded to 94 square kilometers. Zhangjiang thus became a new city in Shanghai.

Zhangjiang's master plan proposes an urban strategy to overcome its relatively isolated status as an independent zone and integrate the development with Shanghai. The density continues to rise, and urban amenities are further developed. Receiving the designation of a science city, Zhangjiang received support from the central government to build an extensive research infrastructure, such as the Shanghai Synchrotron Radiation Facility, with a 1.2 billion Yuan investment. In 2022, about 22,000 enterprises were located in Zhangjiang, among which 1,900 were certified 'high-tech enterprises.' About 170 global R&D centers were invested in by multinational companies. The number of national-level technological centers reached 14 in 2019.

The history of Zhangjiang reveals that it has been created as a 'state strategic innovation space' (Zhu et al. 2023). It was initially promoted under industrial policies to take specific innovation functions, attracting global R&D investment. This is an essentially zonal strategy, with a more streamlined government in order to make it business-friendly. The high-tech park was relatively isolated and low-density in nature. Increasingly, the development has been combined with the overall urban development of Shanghai. Under the confluence of industrial and urban development policies, Zhangjiang strived to go beyond a technological and employment node and become a new city. As a sub-center of Shanghai, it aims to help Shanghai achieve its mission of becoming a national innovation center. Zhangjiang's development, therefore, is not a simply local innovation strategy by large tech firms. It is, in fact, an urban development strategy of the local state, here the Shanghai municipal government. Further, Shanghai's urban development and globalization to become the dragonhead of the Yangtze River region are part of the state strategy of China's modernization (Wu 2000a; Wu 2003a). Zhangjiang combines local and associative innovation and multi-scalar state strategies as a techno space.

The Zhangjiang case reveals that innovation policy has shifted from direct S&T programs, such as the Torch program and 863 programs in the 1990s, to more decentralized innovation governance involving local states and market mechanisms in city development. In other words, innovation has become part of urban development in Chinese cities.

Building and branding a system of high-tech parks

The local state often builds more than one concentrated high-tech area and constructs a system of high-tech parks. These high-tech parks are connected through inter-firm production chains with differentiated and specialized functions. For example, Zhangjiang in Shanghai has 22 parks, including the core of Zhangjiang Science City. Similarly, Zhongguancun in Beijing has 16 parks across the metropolitan region. This is the territorialization of innovation in the metro area. With uneven and specialized development, the metropolitan region becomes a regional innovation system. Therefore, the innovation space in China is a group of nested and linked high-tech enclaves distributed across a wider metropolitan region. This territorialization is achieved through strategic planning and institutional reconfiguration rather than the organization of lead firms.

The expansion of Zhangjiang's brand to other places in Shanghai started with the renaming of the Shanghai High-tech Development Zone with a Zhangjiang title in 2006. This extended Zhangjiang's special policy to six parks in Shanghai. The policy hoped to replicate the Zhangjiang model in these parks. Since then, Zhangjiang has no longer been a single spatial entity, but a group of high-tech parks. The renaming helps to enhance Zhangjiang's visibility and create a brand. Although these parks are still managed independently by their own management offices, the reconfiguration exercise helps to strengthen the collaboration rather than competition between them. As a result, the brand of Zhangjiang was extended from 25 square kilometers of one core high-tech park into 47 square kilometers under the system of 'one zone and six parks' in 2006. The number of collaborative parks increased to 12 in 2011, when Zhangjiang was granted the new title of 'National Demonstration Zone of Indigenous Innovation.' At this point the space had been extended to 291 square kilometers. In 2013, the number increased to 18, with a combined space of 488 square kilometers. Eventually, in 2015, the number increased to 22, with the space under the Zhangjiang brand reaching over 500 square kilometers. The expansion of the Zhangjiang brand provides an opportunity to receive policy support from the central government. For urban entrepreneurialism, place promotion usually aims to attract global capital. However, the entrepreneurial behavior of state entrepreneurialism here is more than a reactionary tactic to external investors. It is also an effort to attract more political support and resources from the central government.

For biotechnology, for example, the high-tech system allows each park to build biotechnology according to its strengths. The policy strengthens the linkage of biotech with related industries and develops commercialization bases in the metropolitan region. The complex biotech network across the metropolitan region designates biotech-pharmaceutical R&D and production bases. For example, Zhangjiang High-tech Park has been expanded southwards since 2009, when Pudong annexed the former Nanhui district. This core area focuses on biotech R&D, and the extended area in the south provides space for pharmaceutical industries to strengthen commercialization. In the central area of Shanghai, around several hospitals, medical schools, and research institutes in the Fenglin area of Xuhui district, this area accommodates clinical research, contract research organizations (CROs), and new drug production.

Further out in the metropolitan region, an animal vaccine production base is developed. The three outer suburban districts are production bases for the biotech and pharmaceutical industries. For example, the Fengxian base under the national Torch Program becomes the production base for raw chemical materials and expert-oriented medicine production. The development of these bases reflects the spatial expansion of biotech and related industries in the Shanghai metropolitan region. This system of high-tech parks shows that spaces for innovation—biotech in this case—become urban-based. These spaces are further intertwined through the cross-ownership of parks. For example, Zhangjiang High-tech Development Corporation purchased the majority share of another medical park in former Nanhui, and pharmaceutical in Zhangjiang subsequently implemented commercialization in the medical park.

In short, the spatial dynamics of high-tech parks in general and biotech within these innovation spaces, in particular, show the effect of a multi-scalar state at the high-tech park, district, municipal, and central governments. This section shows how governing innovation spaces operated locally in concrete geographical spaces—in China as a series of nested high-tech parks. Industrial policies for innovation are implemented by constructing the materiality of innovation, or literally 'building' a regional innovation system (Zhang 2015). Instead of focusing on internal knowledge dynamics within the firms and their networks, the next section will examine urban conditions upon which innovations are supported. These conditions include assembling the talent and human resources suitable for R&D, using the land–finance nexus to support infrastructure development, providing financial support, and developing a spatial system of high-tech parks with more collaboration and division of labor.

State strategies: building an innovative nation

This section explains how governing innovation and its space is under the state's innovation and urban strategies. The city is an essential scale at which these strategies are materialized. Hence, innovation is associated with localities as 'local development.' However, a perspective of locality or innovative milieu is insufficient. Instead, as we argued here, the city should be seen as an assemblage where multiple actors negotiate and engage with each other for their interests. Understanding the post-reform political economy, especially the multi-scalar governance of the Chinese state and its interaction with globalization, helps to reveal why high-tech parks are the localities where governing innovation is performed.

These innovation spaces are not a loose geographical concept, in contrast to Silicon Valley. They require the central government's explicit endorsement of their status and spatial boundaries as 'economic and technological development zones' (ETDZs) or national High- and New-Technological Development Zones (NHNTDZs) (Appelbaum et al. 2018). For example, in the Tianfu New Area of Chengdu, a 'Unicorn Island'—now renamed a Tech and Bio Island—has been designed with a master plan by Zaha Hadid Architects (Figure 6.2). These are substantial operational units managed by the government and developed by the development corporation. Hence, they are state spaces. We will detail their operation later. Here, we explain governing under state strategies presents state centrality and planning rationality.[5]

Figure 6.2 The Unicorn Island, located in the Tianfu New Area, Chengdu.

The strategies of the central government for innovation

As discussed in the history of innovation models, China has a long tradition of guiding industrialization. Associated with the priority to develop heavy industries and military capacities during the Cold War, the central government strategically mobilized innovation in key defense industries through assembling resources. Economic decentralization and foreign investment under the reform and 'open-door' policies led to a decline in the role of the central government. Local governments became primary organizers of economic development. The central government only initiated and maintained some critical funding programs, leaving the innovation task to enterprises, in a process known as enterprises-centered innovation (Liu and White 2001). However, the lack of overall innovation drove the central government to publish the MLP in 2006, the first explicit strategy to transform China from the world's factory to the world's innovator (Appelbaum et al. 2018).

The MLP advanced the strategy of 'indigenous innovation.' The notion of 'indigenous' means ownership of IP, which could be created through diverse market routes. This strategy reveals the association of innovation and nation-state, which is less exceptional to China, considering the history of mission-oriented S&T in the rise of Silicon Valley. The mega-projects proposed in MLP are funded mainly through the financial market using a financialization approach—as discussed in Chapter 3, which has been made possible by credit expansion following the global financial crisis. With the rising financial capacities of the state in urban development, the central government not only endorses the special status for high-tech parks but also allocates resources for key science facilities through partnerships with local governments. For example, the development of the high-tech park in Chengdu's Tianfu New District is perceived and associated with the national strategy of indigenous innovation (Figure 6.3).

In 2015, China launched 'Made in China 2025', an explicit industrial policy similar to Germany's Industry 4.0. The strategy intends to develop advanced manufacturing industries as China faces a shortage of cheap labor. The strategy must promote innovation, as China lacked the technologies to move up the value chain. Hence, innovation has been an overwhelming target, with increasing funding in R&D. The strategy focuses on digital and smart technologies. The subsequent 13th Five-Year Plan from 2016 continued to implement the strategy. The 'Made in China' indicates innovation as a state strategy.

While the strategy emphasizes the importance of innovation, the actual implementation has been flexible. Domestic innovation and production capacities are the targets of various implementation plans, continuing into the 14th Five-Year Plan. The state-led innovation initiatives created the space rather than closing down international collaborations. Besides the challenges of identifying new industries and emerging technologies, the central state leaves local states to seek opportunities and compete with each other for market realization, as shown in new energy vehicles. One of these strategies is to designate key innovation spaces. Recent geopolitical tension has further strengthened the importance of state innovation strategies but also brought uncertainties and flexible approaches to implementation.

Figure 6.3 The development of the high-tech park, the Unicorn Island in Chengdu and its association with the national innovation strategy.

The entrepreneurial local governments in building innovation spaces

Economic decentralization and the tax-sharing system incentivized the local government into entrepreneurial agents. A market approach to urban development centered on 'land finance' has further enhanced their capacities. Under the strategic guidance of the central government to foster 'indigenous innovation,' they initiated policies to develop a defined space of innovation; for example, Shanghai initiated the policy of 'focusing on Zhangjiang.' The Chinese approach is to 'build' innovation through the high-tech park. The management committee coordinates the development, which unifies various fragmented government departments above the park. It overcomes bureaucratic constraints and is 'entrepreneurial,' administratively streamlined, and business-friendly. The development corporation carries out the actual construction of park infrastructure through land development and market means. The municipal and district government and park quasi-government work together to 'build' innovation spaces. They foster linkages between universities and industries, even relocate state research institutes and universities into the high-tech park, and encourage the development of joint ventures between universities and industries. One important policy is attracting talent, which will be explained below.

Attracting talent through the multi-scalar state

Returned entrepreneurs play an important role in the success of high-tech development in East Asia. For example, the returnees from Silicon Valley to Hsingchu in Taiwan constituted a critical link between these two locations. This has been a decisive factor in the development of the science park (Hsu and Saxenian 2000). Overseas returnees significantly enhanced the scientific research capacities in China, as many became academicians at the Chinese Academy of Sciences and the Chinese Academy of Engineering. The returned entrepreneurs also bring business management and financial knowledge.

For the 'creative classes,' it is suggested that the tolerant local culture and cultural diversity are attractive for them to concentrate in some large cities (Florida 2002). Creating urban amenities such as green spaces and cultural facilities helps to attract talent. The synergy between technology, talent, and tolerance is used tactically to foster creative industries and innovation (Florida 2002). In China, the overseas returnees are understandably concentrated in first-tier cities such as Shanghai, Beijing, Guangzhou, and Shenzhen because of their high-quality living environment and vibrant local economies. While the notion of 'creative classes' suggests a perspective of a job following people, the Chinese approach focuses more on creating the business environment to attract investment—'building the nest to attract phoenix.' One primary reason for overseas returnees to come to these large cities is that the science parks built in these cities provide various supports for their start-ups.

Attracting overseas talent is also a state strategy. For example, the central government initiated the 'Thousands of Talents' program in 2009, explicitly targeting returnees in the high-tech industry, finance, and other knowledge sectors. The provincial and local governments set up their programs to compete with each other to attract talent, as they do for foreign investment. The talent concentration is an outcome of complex multi-scalar governance processes. Municipal governments strive to attract talent through inter-city competition. While the central government initiates the talent program, local governments operate these programs and develop them as local policies to strengthen their advantage in talent attraction. For the centrally operated programs, the local governments must bid for these posts subsidized by the central government. The top universities and prominent research institutes are the hosts for the overseas returnees of scientists. Because these top universities are located in large cities, more talents eventually settle down in these large cities. The local governments with more substantial financial capacities can afford more generous conditions and posts. As a result, the talent concentrated in the first-tier cities and provincial capitals. In addition to the high education sector, research institutions and industrial businesses also manage to attract talent. The science park, with the concentration of high-tech enterprises, performs strongly in attracting talent. For example, Zhangjiang in Shanghai hosted more than half of the returnees and one-third of the newly founded enterprises by returnees in Shanghai in the 2010s.

The returnees also bring much-needed business knowledge and help to fill a gap between scientific research and industry. While China has many low-cost scientists and researchers, the lack of business knowledge is a constraint. Successful returnees have experience managing research or marketing in industries like 'big Pharma' for biotech. They identify business opportunities and develop products based on their patents (Zhang 2015).

In innovation research, significant attention has been paid to the consumption and cultural environment of the city, while the effect of government policies is not sufficiently explored. In the case of Chinese high-tech development, it is more than local culture and amenities. All these may have an effect, but talent programs directly affect the labor force. These programs are national strategies and initiatives implemented through the multi-scalar state and interactions through cultural affinity and social relations (*guanxi*). The 'technology, talent, and tolerance' formula is conceived in a more neoliberal and entrepreneurial governance environment. Under state entrepreneurialism, the multi-scalar state assembles the critical human capital component for innovation through its human resource policy and talent programs.

The central state talent program helps to attract overseas returnees. The program does not specify the location for returnees to settle. Talent distribution depends on the available innovation spaces and local conditions. Cultural amenities and an open and entrepreneurial environment (such as Shenzhen, a Special Economic Zone) certainly matter. The availability of innovation spaces such as top-rank universities, major research institutes, and science parks, as well as facilities and infrastructure and matching funds by local governments or institutes, are all crucial factors. The local governments use the high-tech park to strengthen their applications for the talent post funded by the central government. This aggregates the imbalance of innovation capacities across the hierarchy of cities. The first-tier cities capture more talents with better universities, working conditions, and educational facilities for their children. In short, talent competition is a more direct outcome of government policies than those factors for labor mobility, which are generally applicable to all cities, and China is no exception.

Market operation

The development corporation

While governing innovation spaces is covered under state strategies, the implementation is achieved through market actors and operations. Four large state-owned corporations have developed the Pudong new district, of which the Zhangjiang Group led the high-tech park development. The governance approach adopts a 'big market, small government' model (Zhu et al. 2023). As a state industrial corporation, the Zhangjiang Group is owned by state assets and the State-owned Assets Supervision and Administration Commission (SASAC) of the Pudong new district. The Zhangjiang Group owns the land in Zhangjiang as its asset. Through its subsidiaries, the group also organizes industrial production. One of the subsidiaries was the Zhangjiang High-Tech Park Development Corporation Limited (the Zhangjiang Development Corporation hereafter, now under the SASAC). It plays a role in park development, similar to urban development corporations or *chengtou* (Feng et al. 2022a).

Based on the asset of the integrated circuit industrial port in Zhangjiang developed by the Zhangjiang Group, the development corporation was listed on the Shanghai Stock Exchange in 1999 and thus raised finance from the capital market. Like *chengtou*, the corporation does not directly engage in the process of innovation but provides infrastructure for the Science City. It is a high-tech property manager. With the status of a listed company and development corporation, the corporation also acts as a local government financial vehicle (LGFV), combining governing technological innovation and land finance. From an optimistic view, the LGFV mobilizes capital to finance infrastructure for innovation activities and even directly acts as a fund, for example, through setting up venture capital or the government-guided investment fund as the fund of funds (Pan et al. 2021). There are some successful cases in China, such as the strategic investment in BOE Technology Group in Hefei. However, most are speculative as the government has limited knowledge of high-tech development.

However, in reality, the market operation by the development corporation often deviates from its original intention to promote innovation to pursue property profits. A significant proportion of the income of the Zhangjiang high-tech development corporation came from property sales after the global financial crisis when the stimulus package led to a property boom. The original purpose was to use the sale income as a subsidy for its business as a high-tech park manager. A similar observation is from the Tsinghua enterprise–LGFV partnership, which profits from selling or leasing properties by building science parks and incubators (Zhang and Lan 2023, p. 14).

Along with the deepening of financialization in the 2010s, the development corporation played a significant role as a financial vehicle in addition to a property developer. During the 2020 pandemic, the net profit of the development corporation was entirely from its venture subsidiary. Despite a rapid increase in high-tech firms in Zhangjiang from 2016 to 2019, the annual revenue only increased by 5.0 percent. This indicates that the development corporation, as classified by the government as a functional type of corporation that is measured by its achievement of the assigned task rather than profitability, follows more plan rationality than market rationality. Regarding direct profitability measures to the corporation, its high-tech management business is relatively 'unsuccessful.' However, the development of high-tech firms led to increased taxes to the local government—here, the Zhangjiang Science City office acts as a local government.

To sum up, the state does not solely or directly implement financing high-tech parks and innovation. In this sense, it differs from the state-centered innovation model in the 1950s to 1980s. Instead, the state uses various corporations and financial vehicles to realize its development strategies.

Land-driven development

To achieve its mission of fostering 'indigenous innovation,' the local state deploys market-oriented land development and provides infrastructure for innovation. The Chinese approach combines the development of science parks and innovation clustering—often known as policy-driven science park development. In other words, the cluster is achieved by building the park rather than resulting from spontaneous firm agglomeration. As such, the 'cluster' often has a defined geographical boundary of 'zone' rather than a general geographical notion such as Silicon Valley.

As discussed in Chapter 1 about multi-scalar governance, the local state is not just an agency of the central state. The local state gains some autonomy in economic decision-making through economic decentralization and tax-sharing. The development zone, in general, and the high-tech park, in particular, have more discretionary power and adopted entrepreneurial governance—attracting foreign investment and global R&D centers. The development zone can act more entrepreneurially because the state, through its institutional environment, both formal and informal, permits and drives these zones to do so.

The high-tech park often enjoys special policies endorsed by the upper-level government. For example, Zhangjiang High-tech Park in Shanghai has enjoyed a streamlined administrative approval procedure since 2011. Through land development, the local state gains profits and tax revenue. As such, the local state is interested in land and innovation. To fulfill the mission of innovation-oriented growth, the local state may use the profit from land to subsidize innovation. However, it can act in the opposite direction, namely, using the centrally endorsed mission and innovation activities to justify property development and make a profit.

Because of this close association between land and innovation policy in China, land and infrastructure development has become a priority for high-tech parks. As discussed in Chapter 3 about financialization, the local state used land as collateral to obtain development finance. In practice, the development corporation operates land development and may gain profit from property sales. It can also gain property rentals from firms in the high-tech park. By obtaining profit from land development, the high-tech park may invest in venture capital and, in turn, emerging high-tech firms.

For example, in Zhangjiang, in the early years of development, the government supported the high-tech park through public revenue. Instead of directly investing in the development corporation, the government rented standard factory buildings from the development corporation to provide the high-tech firms. With the revenue income from rental, the development corporation managed to raise more capital from the capital market. With the status of a government-backed development corporation (actually a state-owned enterprise), the development corporation managed to issue a bond with lower interest rates, saving a significant amount of interest.

Hence, land development and the development of high-tech industries are closely linked. For example, in Zhangjiang, the development corporation has a 'land and share swap agreement' to support start-ups. It waives land and property rentals in exchange for a share of the enterprise. When the start-up is developed, the government may transfer the share to financial investors or gain capital when the enterprises get into an initial public offering (IPO).

Multinationals' R&D centers

Multinationals select large Chinese cities to build their R&D centers, which are important drivers for innovation in China. They concentrate on the high-tech parks in global cities for various reasons. Big pharmaceutical companies' internationalization of R&D in Shanghai is not driven by tapping into the local knowledge base. Instead, they aim to access the large Chinese market by developing offshoring activities for early drug development (Grimes and Miozzo 2015). The low labor cost of scientists and lab technicians is an attractive factor. Market access and cost savings are primary considerations for pharmaceutical production rather than local inter-firm networks. Their relocation and development of global R&D centers are similar to the earlier stage of foreign investment in China's development zones because of low-cost migrant workers and China's market potential.

For biotech, global R&D centers are developed by multinationals in association with the relocation of pharmaceutical production as the 'pharma-biotech complex' (Cooke 2004; Zeller 2010). They use the Chinese market to develop new drugs. To them, Shanghai presents a new opportunity as the gateway to the Chinese market. The high-tech park provides material support for relocating its R&D centers within the city. For drug development, multinationals collaborate with Chinese biotech companies. The latter are contract research organizations (CROs) that offer a price-competitive research service for multinationals. The CROs take research services with a fixed payment and have low risk. However, they do not have the IP and are not interested in the research lines. The available CROs are attractive to multinationals. Many multinational CROs also set up their offices in Zhangjiang, and local-based CROs also managed to gain IPOs in the US and are embedded into the global pharmaceutical service networks.

The relocation of global R&D centers and the emergence of CROs in biotech are associated with many factors available in the Chinese global cities, other than local buzz —frequent and unplanned face-to-face contacts—and existing knowledge dynamics in these localities. These factors happen to present at the urban scale, more specifically in the high-tech parks of these cities. As discussed earlier, these factors are created through conscious actions of city-building. Under state strategies, these factors are assembled at these localities through market developments, such as constructing high-tech parks. The case is significant, as Zhangjiang was a non-urban location, later turning into a peri-urban development zone as Shanghai developed the Pudong New District and is now moving toward the Science City. It is not an urban or local buzz model.

Critical reflection

This chapter describes existing practices governing innovation in China rather than a theoretical or normative innovation governance model. It does not depict a rosy picture of the exemplar of innovation governance. Hence, the model does not mean a successful approach. Instead, it highlights the limitation of entrepreneurial governance when applied to high-tech development and innovation, as these emerging technological sectors are investment-intensive and highly uncertain in financial return. Unlike infrastructure development, they rely on networks and interactions of diverse actors.

For Zhangjiang, we do not present it as a model of best practice, which is often implied in the Silicon Valley model. Zhangjiang's experience might not be replicated elsewhere in Chinese cities. Compared with Shenzhen—China's innovative city—Shanghai is often said to be less innovative. The performance of China's innovation policy is still subject to critical evaluation. Despite some visible concentration of innovation resources, China still faces formidable challenges in developing core technologies, such as integrated circuit manufacturing.

China's governance of innovation through a significant state role is no exception—even in the so-called liberal market model, the 'mission-oriented' entrepreneurial state is presented as a material reality and a prescriptive model (Mazzucato 2018). On the other hand, the Chinese mission-oriented state-centered innovation model has been transformed with the introduction of the market and globalization. It is no longer a state-funded innovation model, though the state continues to play a strategic role. Moreover, Chinese experience suggests that governing innovation is not limited to the S&T sphere, involving the political economy of urban development and governance.

We must understand the political economy and the processes that contribute to developing innovation capacities (Zhang and Wu 2019). In this chapter, we show that China's innovation capacities are highly uneven and concentrated in the high-tech parks of its global cities. This is not due to the 'nature' of the city as an agglomerative space (Scott and Storper 2015). Instead, the city is an assemblage where multiple agencies—the multi-scalar state and lead firms in GPN—interact and affect innovation in the cities. They act upon the conditions for knowledge generation. These conditions are closely related to the concentration of innovation and uneven development. The intrinsic dynamics of knowledge generation—agglomeration, encountering, and diversity—are necessary but insufficient to understand the spatial concentration of innovation. In other words, uneven development is not simply derived from knowledge dynamics that highly value face-to-face interactions. The chapter explains how such conditions occur rather than a given feature.

Just as the available capital itself does not create surplus value—the surplus value is created in labor commodification by capital and the laboring process in capitalist production, the geographies of clustering, agglomeration, and diversity do not directly create innovation. The actors manipulate these conditions in the urban development process—or, more theoretically, noted as the 'urban process' by David Harvey (1978)—contributing to the generation and concentration of innovation. The multi-scalar state governs innovation by creating and maintaining the necessary conditions for innovation. The state itself does *not* create innovation. Instead, it shapes and affects innovation through city-building (Zhu et al. 2024). In this sense, economic development and innovation under state entrepreneurialism do not mean a centralized state fulfills every task. It is different from the developmental state thesis, which has been duly criticized by the relational perspective of a global production network (Yeung 2019b).

Hence, as we have shown in this chapter, to explain why China has recently developed its innovation capacities and why innovation concentrates in its large cities and, in particular, the high-tech parks, innovation centers, and the 'science cities,' we need to identify the real actors and their agencies in the city rather than resorting to a general notion of agglomeration, diversity, or relatedness between economic sectors. The recent cultural and institutional turn significantly expands the scope of the new economic geography beyond economic analysis. However, the attention to urban development and its political economy dynamics remains insufficient.

We need to understand the political economy process—city-building and urban development, in which innovation is part of the parcel and the associated politics of governing innovation (Zhang 2015).[6] GPN helps to reveal this process as a process of knowledge and economic production, paying attention to firms. A dynamic GPN perspective further adds the evolutionary aspect of GPN change, and 'strategic coupling' and de-coupling introduce an institutional understanding of the relationship between firms and regional assets, which helps to reveal the political and governance aspects. They provide insightful perspectives and valuable analytical frameworks. They can be applied, as they are flexible—regarding the nature of innovation governance in China.

Here, to address this 'nature,' we provide a more concrete political economy analysis of innovation in China, which, similar to studies of innovation in other places in the world, has so far been explained extensively through economic factors, production processes, and a probably quite generic institution. This chapter highlights the particular institution or governance feature—a hybrid and 'national indigenous' model of innovation, combining some selective features and components of all three varieties—the entrepreneurial, associative, and developmental models. These combinational features—state strategic centrality and market operations—lead to what we call 'state entrepreneurialism,' China's salient innovation governance model.

Notes

1 See also Zhou (2008b). This chapter, however, examines innovation in the context of national and local development. It regards innovation more than knowledge production and understands how knowledge production is made through economic development. Hence, the national indigenous innovation model is an economic development model.

2 For state capitalism, see Alami and Dixon (2020) and Alami et al. (2021). The role of the state is also discussed in this late capitalist context, see Moisio and Rossi (2020).
3 According to actor–network theory, the network itself could be regarded as an agency; the network discussed in economic geography is different from an assemblage (McFarlane 2011) or a 'regional assemblage' of governance beyond defined localities (Allen and Cochrane 2007).
4 They are the founder of RIS, Phil Cooke, and extensive application in Nordic countries, Bojor Asheim, and the founder of related variety and evolution theorist, Ron Bosman. See their joint paper, Asheim et al. (2011).
5 This is similar to the developmental state which is said to consider plan rationality instead of just market signals.
6 This perspective stresses the actual process of building innovation spaces, see also Luo and Shen (2022).

7 Environment

The socio-ecological fix under ecological civilization

Introduction

China faces immense environmental challenges. These challenges result from, and are associated with, rapid economic growth and continuing urbanization. The impact of Chinese urbanization on climate change is now as much a global as a local issue, as China surpassed the US as the world's largest CO_2 emitter in 2007.[1]

Besides carbon emissions, China faces problems with energy inefficiency, water and air pollution, traffic congestion, and the loss of arable land. The level of PM2.5 was frequently way above the safe standard by the World Health Organization.[2] Environmental incidents had disastrous impacts on public health. For example, Tai Lake experienced a severe algae bloom problem—eutrophication—in 2007. Every day during the outbreak the workers had to clean the algae manually. When global e-waste is received, the recycling industries in small cities such as Guiyu in Guangdong cause heavy metal poisoning and irreversible pollution (Wang et al. 2021; Lora-Wainwright 2017).

Under the growth-first strategy in the post-reform era, local governments pursued economic growth at a high environmental cost. Despite the central government strengthening pollution control, the local government did not forcibly implement environmental management. Rather, they often used the opportunities of environmental initiatives such as eco-cities to promote local economic growth and land development.

Since 2012, the central government has strengthened environmental governance through a mandate of 'ecological civilization.' It began to impose a stricter requirement for environmental protection and, later, climate change. This is often described as 'authoritarian environmentalism.' Because of the multi-scalar nature of the Chinese state, the tension between economic motivation and ecological civilization persists. More fundamentally, the dynamic relation between capital accumulation and territorial logic underlies such tension. However, China is not exceptional in this regard. Such a contradiction is widely prevalent in state capitalism, as shown in Chapter 1. Nevertheless, China's state entrepreneurialism does offer a visible presentation of the tension between economic growth and environmental governance (Zhang and Wu 2023).

China's environmental governance has been transformed in response to the environmental crisis. The administrative system managing the environment has been re-configured, and rationality has transformed. Governmentality—political rationalities and governmental technologies—has experienced a significant shift from a growth-oriented one to the wish for human–environment civilization (Zhang 2021, 2023). This chapter examines this shift—a change beyond rhetoric—along with the associated socio-ecological fixes and their implications for our critical understanding of environmental governance. The fix is used metaphorically to fix an environmental problem,

DOI: 10.4324/9781003355175-8

but it is a theoretical concept associated with capital circulation and state regulation, or even specifically as a governmental technique.

Theoretical perspectives on environmental governance

The socio-ecological fix

The concept of the socio-ecological fix is derived from the general literature on the 'spatial fix' in advanced capitalism. The spatial fix, originally developed by David Harvey (1982), means to fix capital over-accumulation through spatial expansion. It is achieved by shifting capital from the primary production circuit to the built environment's secondary circuit. As a result, the capital is absorbed into the built environment. The spatial fix, however, is only a temporary solution, and hence, in essence, it is a 'fix' rather than a permanent dissolution of the fundamental contradiction inherent in the capitalist mode of production.

Applying the spatial fix to the natural environment aspect, the fix notion is further developed into the 'sustainability fix.' The concept here describes the selective incorporation of environmental goals into urban governance (While et al., 2004). The purpose of developing environmental projects is to fix the crisis of capital accumulation. It is not necessarily a shift of capital from an over-accumulated production sphere to environmental projects. Environmental development provides the necessary conditions for the capitalist economy. It is imperative to fix the environment in order to sustain capital accumulation.

Therefore, the sustainability fix differs from what is commonly known as 'greenwashing,' since the former does actually attempt to improve the environment. However, these environmental maneuvers are only temporary solutions; they do not address the substantial causes of environmental challenges. Environmental policies are formulated through policy mobility to promote 'best practices' (Témenos and McCann 2012). These policies intend to support urban development and deal with the economic crisis rather than broader societal and ecological problems.

Further, the fix notion is extended into the socio-ecological fix (Ekers and Prudham, 2017). The fix is not limited to addressing the crisis of capital accumulation. Unlike the previous sustainability fix, the socio-ecological fix is not driven by the imperative of economic growth for the 'production of space.' Instead, it intends to support the 'production of nature.' The fix advances ecological changes. The notion of the socio-ecological fix goes beyond the political economy, but it does include political ecology. The fix is about modifying the environment through changing socio-nature relations.

While still adhering to capitalism's political economy and ecology, upgrading the ecological environment is an attempt to cope with social reproduction and nature crises beyond capital over-accumulation. The concept is not limited to urban development but stresses environmental objectives through environmental regulation. In other words, the fix is not simply for an economic purpose. Public investment may be mobilized to upgrade the environment (Castree and Christophers 2015). The socio-ecological fix requires more substantial institutional changes.

Within the overwhelming environmental target, the state may be subject to restructuring, hence we introduce a new concept of 'eco-state restructuring' (While et al. 2010), which will be explained later.

In sum, the concept of the socio-ecological fix, as can be seen from the history of the spatial fix in general, is conceived within the crisis-response framework. The fix refers to a solution to the crisis under capitalism. While the original notion of spatial fix or an extended version of the sustainability fix considering the environmental aspect as adhering to capital accumulation, the socio-ecological fix intends to solve a broader crisis of capitalism in social reproduction and ecological sustainability. These fix notions reveal that there are various motivations for applying the fix. These motivations tend to be associated with capital accumulation. They should be investigated more in a concrete geographical context. The space in their notions is more generic, as different localities are all under capitalism, representing the variegated practices. The motivation is less investigated in territorial politics—it is less clear why and how the actors invent and deploy the socio-ecological fix. To understand the motivation, we must examine the multi-scalar state in environmental governance.

Eco-state restructuring

The concept of 'eco-state restructuring' was proposed by While et al. (2010) to describe the changing regulation of economy–environment relationships at the urban and regional scales. It is defined 'as the reorganization of state powers, capacities, regulations, and territorial structures around institutional pathways and strategic projects, which are (at least from the vantage of state interests at a given moment intimate) viewed as less environmentally damaging than previous trajectories' (While et al. 2010, p. 80). They described three waves of eco-state restructuring in the world. The regulatory objective was pollution control from the mid-1960s to the late 1980s. From the mid-1980s to the mid-2000s, sustainable developmentalism and ecological modernization rose to become the dominant rationality. From the late 1990s, carbon control began to occupy state capitalism's centrality.

The concept highlights the recent dominant consideration over carbon control to deliver a low-carbon future. It stresses territorial politics, the necessity to re-regulate society and space, and carbon governmentality. In contrast to neoliberal governance, which allows competitive states to trade off environmental and economic goals, the recent eco-state restructuring is geared toward strengthening territorial-based targets and trading. The concept also situates the state regulatory system in the global geopolitics around carbon control and climate change. It effectively points out a distinctive new agenda for state environmental regulation, which has notable implications for territorial governance.

This eco-state restructuring concept reveals a similar change in China's environmental governance from a growth-oriented polity with ineffective pollution control to a strengthened rationality of ecological civilization. It also highlights two important aspects of environmental governance. First is the state's role in environmental regulation that moves beyond neoliberal governance, and, second, this rising agenda is mainly due to carbon control as an overwhelming problem for urban and regional governance. As for the second aspect, China faces the same carbon control imperative and geopolitics, which are broadly perceived as threats to its civilization. The perspective of eco-state restructuring thus points out a possibility of environmental hegemony, similar to authoritarian environmentalism but simultaneously with the more multi-scalar state acting across urban and regional territories.

Environmental transition and the multi-level perspective

Related to environmental transition through green innovation is the multi-level perspective (MLP). Focusing on the socio-technical transition, this is an innovation rather than a governance model. Initially developed by Geels (2002), and with a recent elaboration (Geels et al. 2014), the perspective has been extensively applied to studying the tradition of urban sustainability (Bulkeley et al. 2014). The perspective applied in the sustainability transition explains how green innovation becomes mainstream. This process involves the interaction between different levels: emerging 'niches' of innovation, socio-technical 'regimes' preserving the status quo, and slow-changing exogenous 'landscape' factors. The model has also been applied to Chinese environmental transition (Yu and Gibbs 2018; Westman et al. 2019; Huang et al. 2021; Zhang et al. 2021).

The multiple levels are conceptual or analytical layers. First, innovations initially emerge in a limited scope, called the niche. They are subject to sociology-technical rules at a broader scope, called the regime. Nevertheless, the regime is situated in an even broader landscape of social and physical conditions. The landscape factors, or these contextual conditions, can change and create external shocks and windows of opportunity for niche innovation to expand and become mainstream technologies.

In geographical studies, there are efforts to introduce political agency into this description of the material narrative of the process (Lawhon and Murphy 2012). Here, explicit attention to power is needed. The notion, probably from an implicit background of liberal market economies, regards the local 'market' innovation as a bottom-up process, driving the socio-technological transition. To a lesser extent, the developmental state is considered in the socio-technological transition (see Chapter 6 on innovation). Even at the municipal or local level, the 'place-based leadership' from economic geography literature highlights the role of localities (Hu and Hassink 2017). As will be seen later, when we discuss China's eco-cities and low-carbon innovations, the municipal government, as the local state, plays a significant leadership role.

As explained, there are some criticisms of the MLP. First, the terms niche, regime, and landscape do not necessarily suggest an agency in themselves. In China's case, the multi-scalar state is underestimated in this perspective. The notion of multi-level innovation is different from multi-scalar governance. Second, the tendency of the regime to maintain the status quo might not be universal. Mapping the level with the municipal government as the urban-level gatekeeper, it has vested interests in land finance and maintains the existing regime. However, it is also affected by the national political mandate at a higher level. It can also be a niche advocator in response to pressure from the central government and external 'landscape' factors. The combination between MLP and the multi-scalar state might be fruitful (Zhang et al. 2021).

Despite these limitations, the MLP is still helpful because it brings a much-needed bottom-up perspective involving multi-level interactions to understanding China's environmental governance. In a different stream of literature, China is noted for its policy experiments (Heilmann 2008). The concept has also been applied to China's environmental governance (Lo et al. 2020; Lo 2023; Xie et al. 2023). Niche innovation occurs in its institution upon the changes in political economy conditions ('external shock') and is endorsed by the central government (a broader national 'technological regime').

The history of environmental governance in China

China's environmental governance has experienced three major shifts, corresponding to the history of economic development. From the 1950s to the 1970s, the state under the Mao era adopted a modernization approach to modify the environment, disregarding ecological principles. The environmental engineering campaigns included the elimination of sparrows as one of the 'four pests,' destroying natural forests to fuel steel furnaces, and filling wetlands and lakes to expand grain production in the 1950s. The relocation of industrial enterprises in the Cold War from coastal areas to inland regions severely impacted the natural environment in these remote and rural areas (Shapiro 2016). China's environmental management led to other policy priorities, such as grain production, steel production, and industrialization during this first period. The awareness of environmental protection was absent. The negative impact on the environment has a long legacy and poses constraint for China's development.

From the 1980s to the 2000s, economic development became a priority in the second period under Deng Xiaoping's policy of pragmatism. Pursuing profits under economic devolution further created many environmental problems, including air pollution, water shortage and flooding, rural land encroachment, loss of green spaces, and public health crises like 'cancer villages' (Lora-Wainwright 2017). The environmental crisis led to strengthened environmental policies and the restoration of environmental authorities. However, in this period, the introduction of market mechanisms and the desire to get rich first led to the excessive exploitation of natural resources, illegal waste dumping, and pollution discharge. In coalition with the private entrepreneurs, the local governments were tolerant toward polluting industries. The underdeveloped local economies, precarious living conditions, and environmental ignorance led to an e-waste disassembling industry in Guiyu, southern Guangdong (Wang et al. 2021).

The environmental problems raised middle-class awareness of environmental protection and led to public protests over environmental incidents. China's integration with the global economy in the 2000s worsened the impacts on rural land, air pollution, and the rapid increase in carbon dioxide emissions. China's environmental issues have also become global and are linked to the geopolitics of climate change. In the second stage, environmental governance is characterized by an effort to manage environmental crises with enormous implementation challenges.

From the mid-2000s to the present, the state took serious actions toward greater environmental control in the third stage. Tackling severe air pollution became a priority for the 'green' Beijing Olympic Games in 2008, and, after that, there was a continuing effort to relocate polluting industries from Beijing to nearby Hebei province (W. Wang et al. 2023b). At the central government level, the environmental protection administration has been strengthened by upgrading the previous environmental bureau to the Ministry of Ecology and Environment and a newly renamed Ministry of Natural Resources to secure land protection. Most importantly, in the late period of Hu Jintao and Wen Jiabao, 'ecological civilization' was coined following a 'scientific approach to development.'

The fourth stage began in 2012 when President Xi Jinping explicitly set the agenda of 'ecological civilization.' This has become the dominant ideology in environmental governance. His well-known slogan—'clear waters and mountains are as valuable as mountains of gold and silver'—aims to shift the mentality from growth first to 'better first and then faster' with an emphasis on environmental quality. The 'National New Urbanization Plan' (2014–2020) prioritizes two policy agendas: the implementation of ecological civilization; and migrant social integration for a 'harmonious society.' Green GDP has been introduced to evaluate local governments as the priority shifts in this period.

The fourth stage of environmental governance illustrates the re-centralization of environmental management (Kostka and Zhang 2018). However, environmental campaigns, such as the restoration of grasslands, resulted in the relocation of minorities in the western region and the transformation of local ecosystems, raising issues of environmental justice (Yeh 2009; Shapiro 2016; Rodenbiker 2022). With increasing tension over global climate changes, China shifted from emphasizing development opportunities in developing countries to viewing it as a global threat to the Anthropocene. Even with President Donald Trump's withdrawal from the treaty in 2017, China reasserted its commitment (Li and Shapiro 2020). In 2020, China announced its ambitious objective in dealing with climate change: to hit peak CO_2 emissions

Figure 7.1 The water treatment facilities near the Erhai Lake, the city of Dali in Yunnan province.

before 2030 and to achieve carbon neutrality by 2060. The announcement demonstrates its wish to take a significant step in the fight against global climate change.

In the third and fourth stages, Chinese cities witnessed various environmental experiments,[3] such as eco-cities to foster green urbanism and eco-innovations and the greenway as a green infrastructure for better environmental quality and amenities. For example, in the tourist city of Dali, the protection of Erhai Lake has become a priority in urban governance. Along the lake, an ecological corridor has been built with support from the central and provincial governments. The water treatment facilities in the region have been built to stop the inflow of untreated water into the lake (Figure 7.1).

The following sections explain these developments from the perspective of the 'socio-ecological fix' in its environmental governance.

Eco-cities and low-carbon development

Like the fever of establishing development zones in the 1990s, the frenzy of eco-city building became widespread in the 2000s. By 2011, more than 259 cities in China had proposed to develop eco-cities. The central government promoted the initiative of eco-cities. Later, with a rising agenda of low-carbon economies, many eco-cities tried to integrate the low-carbon policy to become 'low-carbon eco-cities.' In contrast to the original eco-city concept (Richard Register 1987), which means the application of ecological principles to city planning, design, and management, the claimed eco-cities are largely newly built zones in the periphery of metropolitan areas. This is similar to new developments such as Masdar (Cugurullo 2021).

China first attempted to build an eco-city in Dongan at Chongming Island of Shanghai (Wu 2012, 2015b; Chang and Sheppard 2013; Chang et al. 2016). This was claimed to become the world's first eco-city. The experiment was hoped to be replicated later in China and the rest of the world. Dongtan would be a new city, spread over an area of 84 square kilometers. The project led to further experiments in Wuxi and Tianjin, which were built. Dongtan planned to build a new city with a population of 10,000 by 2010 in order to showcase it at the World Expo; this was expected to reach 80,000 by 2020, before eventually reaching half a million. Shanghai Industrial Investment Corporation, a large state-owned enterprise under the municipal government, owns the land. Arup, a London-headquartered engineering consultant firm, was commissioned to provide a master design. Building an eco-city came from Arup's notion of 'integrated urbanism.'

With an initial planned population of 10,000 in a vast space, the development was conceived as being a low-impact neighborhood. But it is close to the reserved wetland area, and the project failed to materialize. Besides the known issue of land quota and the jailed mayor who supported the project, the key issue is that the location is too far away from Shanghai to make it a viable real estate project. The project did not manage to attract investment. Arup is a consultant firm rather than an investor. Although there was an attempt to attract HSBC as a potential investor, it did not materialize.

As an ecological experiment, almost all eco-cities resulted from policy mobilities and collaboration between Chinese and overseas developers (Hult 2015; Rapoport and Hult 2017). Dongtan is a collaboration between the Shanghai Municipal Industrial Corporation and Arup based in the UK. The flagship of Sino-Singapore Tianjin Eco-city is a collaboration between the Chinese and Singapore governments. Although endorsed and supported by the two national governments, the project is a joint venture between the Tianjin TEDA investment group and a Singaporean state-owned enterprise. It is located in Tianjin Binhai New District. The project hoped to showcase the 'ecological fix' of unusable saline-alkaline and polluted land. The project also adopted the concept of a neighborhood unit in Singapore. The plan also allocates over 40 percent of land to green spaces. This reflects the Chinese effort to make it more of an ecological experiment than a real estate project. The eco-city also absorbed many eco-technologies.

However, what made the project materialize was its intense real estate developments. The eco-city accommodates many real estate projects and is built through a relatively high-density approach, just like many new towns in China. The eco-city is more or less a residential development rather than a balanced city with job opportunities. Although the original idea was also an ecological industrial park for Singapore enterprises, like a development zone, with environmental planning, the eco-city attracted just over one thousand small businesses without large firms.

Other eco-cities or eco-districts are more oriented toward the development of low-carbon industries. For example, the central government endorsed the new district of Guangming in Shenzhen as a low-carbon city in 2010 (Fu and Zhang 2023). It accommodates LED industrial clusters as a high-tech industrial park with upgraded manufacturing. Learning from Dongtan, Wuxi also planned a low-carbon city as a small part of its Taihu New Town. Although the original intention was a Sino-Swedish collaboration, the project did not materialize because of a lack of investment. As a whole, however, Taihu New Town managed to remove inefficient and polluting town and village enterprises (TVEs), promote economic transformation toward creative industries, develop renewable energy industries, create green space and infrastructure, and promote low-carbon transition. Then, compared with earlier developments oriented toward real estate in the eco-cities, which the central government criticizes as being more development-oriented, the later experiments have been strengthened with an ecological agenda.

The city of Wuxi, for example, developed a hub for PV panel manufacturing and in the 2000s became well-known for producing the world's largest PV panel manufacturing company—Suntech Power Corporation. The city has decarbonized using solar energy in industrial sectors and installed significant solar PV capacities in recent years. In the early 2010s, the Wuxi PV industry reached almost 10 percent of its GDP. The development of solar industries coincided with increasing ecological priorities and the long tradition of local industries. The low-carbon economy is developed as a result of two external shocks: First, the cyanobacteria outbreak and then the collapse of the export market of PV panels. These promoted the municipal government of Wuxi to use the development of Taihu New Town as a base for promoting the domestic use of PV panels and accelerating the energy transition (Zhang et al. 2021).

The Taihu New Town became the national pilot low-carbon eco-city. The support of the central government for solar products is also critical. After the global financial crisis in 2008, the Ministry of Finance started to subsidize the installation of solar PV systems in industrial parks and residential buildings in order to nurture domestic market demand for China's solar industry. Subsidies from the upper-level government were the main initial drivers for expanding distributed PV installation among domestic and industrial users. Although the subsidy was withdrawn later, as the market became increasingly viable, the role of the multi-scalar state in the energy transition in Wuxi is very important. Wuxi's low-carbon economy also demonstrates that while many eco-cities are more or less greenwashing, an ecological innovation agenda has achieved some effect. The local government plays an active role through 'governing by enabling' (Zhang et al. 2021, Yu and Huang 2020).

There are heated critiques on eco-cities. Some senior officials of the central government openly criticized some eco-cities as being fake.[4] Here, we suggest several limitations of Chinese eco-cities. First, from the environmental point of view, Chinese eco-cities are largely imagined as new developments or 'new towns.' It is costly to build these eco-cities through large-scale real estate development. The highway transport infrastructure has to be provided in order to ensure these places are connected and attractive. This also increases the cost of extending the public infrastructure system to a new town. In contrast, old urban areas need retrofit and redevelopment. These newly built eco-cities may not contribute to the environment.

Second, from the economic point of view, eco-cities are viewed as urban laboratories for eco-innovations, where green technologies can be experimented with. However, eco-cities prioritize green economic development rather than protecting the original ecosystem. Places such as Chongming Island experienced ecological engineering and ecological modernization away from nature-based solutions (Xie et al. 2022). The pursuit of urban sustainability and the construction of eco-cities are used to justify pro-growth entrepreneurial policies. These new eco-cities became 'technoburbs', and the development approach is a 'sustainability fix' (Chang et al. 2016).

Third, from a social point of view, eco-cities pay insufficient attention to local communities. Instead, they are aiming at relatively wealthy residents, often leading to eco-gentrification and more significant social-spatial inequalities (Caprotti et al. 2015, Chang 2023). If ever built, eco-cities are large real-estate estates with green buildings and ecological features in enclaves or gated communities.

Building eco-cities has different motivations, depending on the local context and projects. At the national level, there is strong support for economic upgrading, promoting ecological innovation and energy saving, reducing carbon emissions, and improving living standards. There was a stronger interest in economic and real estate development at the local level. Eco-cities often involve international collaboration. Overseas partners wish to gain business opportunities, while the Chinese government hopes to introduce new technologies and investments. These pilot projects are usually designed as replicable models for wider application later.

Nevertheless, in reality, such a policy mobility often fails. There is no appropriate universal standard for eco-cities. The construction of eco-cities reflects the negation of different partners and actors to achieve their interests. Eco-cities are often considered 'best practices' imported from elsewhere through international collaborations. The model of eco-cities is usually created for place promotion and bidding for various global design awards. As such, the model often pays inadequate attention to local contexts. These projects failed because they are not viable urban development projects under the constraint of development conditions, or if they are oriented toward the mainstream practices of land-driven development, they became ordinary real estate projects and failed to achieve sufficient ecological advancement.

The failure of Dongtan did not stop China from experimenting with more socio-ecological fixes. A new eco-island plan has been prepared for Chongming Island. This plan stresses protecting the natural environment and ecological modernization of low-efficient land uses (Xie et al. 2019). Under the strong guidance from the Shanghai municipal government, Chongming is not allowed to initiate large-scale real estate developments. Its polluting industries were closed down. The government insisted that the island should be Shanghai's back garden, reserved for ecological purposes for the sake of the whole city. Similarly, greenways and features are created for environmental amenities and active travel in other places. These new green projects have a stronger ecological focus than eco-cities driven by real estate development.

Greenways: an ecological fix through green infrastructure

China initiated a national campaign to create forests at a regional scale to prevent floods, soil erosion, and sandstorms. However, since the 2000s, greenways have been built in cities and suburbs for landscaping and recreational purposes. The function of urban greenways is to serve as a cosmetic approach to 'city beautification' (Yu et al. 2006). Urban greenways, built as landscaping projects along transportation corridors, have become popular.

The epitome of greenways in China launched the Pearl River Delta (PRD) Greenways in 2010 (Chung et al. 2018). The Guangdong province aimed to build 9,000 kilometers of greenways across cities in the PRD within five years. The ambitious greenway project thus raises a question of its motivation. The literature on greenways in China provides a diverse range of motivations. The initial motivation, proposed as a technical version of green space protection by planners, was quite simple: preventing the encroachment of the built-up area on greenery rural areas. This would be more effectively achieved by introducing recreational uses to the protected area so that the public would monitor and protest the invasion of illegal building activities. Developing greenways adds environmental amenities in peri-urban areas, but there is a shortage of green spaces inside the city.

The construction of greenways, however, is not an urban reforestation project because the project did not plant a large number of trees but rather used existing green spaces along rivers and transport routes to create recreational trails. The greenways connect major parks, nature reserves, scenic spots, heritage sites, and existing rural settlements. The project opens the rural space for urban recreational uses, for example, converting existing agricultural fields and countryside routes for walkers and cyclists (Chung et al. 2018, Zhang et al. 2022).

Despite an initial conceptualization of the greenway for green space protection, creating PRD greenways is, first and foremost, a project for political achievements. It was deemed by the then Party secretary of Guangdong province, Wang Yang, as a political mission: to protect the environment and bring environmental amenities to the people. However, getting benefits for society is not an electoral consideration. This needs to be understood in the context of national strategies toward high-quality development, ecological civilization, a 'harmonious society,' and 'society happiness.' These all together align with the central government's political mandates, contributing to the career advancement of politicians. The greenway was the 'first-order political decree' by the key politician in the base of Guangdong for the next transition to the central government.

In this way, the greenway of PRD should be seen as a 'developmental' project. But this is a new style of development. Compared with industrial zones, large-scale office districts, and new towns, the greenway is a low-cost project using existing green space. The project is never intended to be a real estate project, capturing the value of rural land. In some sections, the project might promote rural tourism and thus benefit farmers.

In another case in Maanshan, a medium-sized city in the less-developed Anhui province, the greenway did not promote rural tourism (Zhang et al. 2020). Most greenway projects are for urban amenities and encourage active travel by building a slow, non-motorized travel system. Thus, the greenways are not just for capturing land values through beautification but are conceived at least as an 'environmentally benign form of urbanization' (Zhang et al. 2020, p. 508). In reality, these two motivations are often mixed, and the balance of these depends upon local actors and specific projects. In Shanghai, along the Huangpu River and Suzhou Creek, a system of greenways has been built. In particular, the Shanghai municipal government strived to open up the shoreline previously occupied by gated communities to the public (Figure 7.2). The project demonstrates a political agenda rather than the motivation of revenue generation through commodity housing development.

For PRD greenways, it is more toward the latter, as post-2008 Guangdong sought high-quality development. The project design, which included 'narrow' trails compared with a grand version of the 'city of parks' of Chengdu, leaned more toward environmental improvement and social benefit projects (Figure 7.3). However, the 'city of parks' combines the greenways with nearby real estate projects and office parks, seeking a synergy between real estate, new high-tech industrial development, shopping centers, and a green environment. It also hopes to attract young professionals with a high-quality living environment.

As intended to be an environmental rather than a real estate project, the PRD greenways demonstrate that China's environmental governance is not exclusively determined by capital accumulation. As seen in other places, such as Taihu New Town in Wuxi, near Shanghai, the adoption of greenways represents the effort to fix social reproduction and ecological crises caused by earlier entrepreneurial profit-driven development (Zhang et al. 2020). It is important to recognize that China's environmental governance includes multiple objectives, not just capital accumulation.

The landmark of greenways is Chengdu Greenways (Zhang et al., 2022) (Figure 7.4). Its urban greenway system is by far the largest in China. Known as the 'greenways of paradise,' the system was launched in 2017 and includes 1,920 kilometers at the metropolitan level, 5,000 at the district level, and 10,000 at the neighborhood level. Further, Chengdu aimed to build the whole city into a park called the 'park-city' (*gongyuan chengshi*) (Zhang and Wu 2024).

Figure 7.2 The greenway along Suzhou Creek in Shanghai, where the shoreline is opened from its occupation by a gated community to the public.

Figure 7.3 The Greenway in Chengdu.

Figure 7.4 The Greenway in Chengdu, combined with nearby commercial developments, at a grand scale.

But whose grand project is this? Before the greenway, Chengdu designated a belt of 500 meters wide along each side of the fourth ring road and seven wedge-shaped green areas as the ecological zones under protection. The ecological zone served as the greenbelt to contain the expansion of Chengdu. In 2012, besides preservation, the then-party leader initiated a project called 'Six Lakes and Eight Wetlands' to introduce landscaping into this preservation area. Very similar to the PRD greenways but in a more aesthetically appealing way, the project introduced recreational uses into the greenbelt, aiming to create a chain of parks on the outskirts of Chengdu. At the time, the urban expansion quickly reached the fourth ring road. The suburbs, however, lacked recreational spaces and amenities. The landscaping project along the ring road was intended for suburban residential development to increase attractiveness. By introducing greenways along the ring road, the 'landscape greenway' no longer achieved the function of urban containment.

However, the lake and park project progressed slowly, lacking investment until its suspension in 2016. In 2017, the new party secretary, Fan Ruiping, arrived. He did not abandon the project. Rather, he relaunched a grand project, the greenway of Chengdu. The project started a new development approach. It re-centralized the power of the municipal government and consolidated land under the control of the municipal development corporation (see later for further explanation). The project captured land value to finance its development and, through the local government financial vehicle, to operate a debt-driven ecological fix.

Within just one year, President Xi Jinping endorsed the effort of green development in Chengdu during his visit to the new district in Tianfu. The title of 'Park City' was officially launched (Zhang and Wu 2024). The following year, Chengdu organized an international forum to publicize this entitlement. It launched a slogan to describe the future of Chengdu as the 'Park City, city of the future.' It is reminiscent of 'place promotion' in Shanghai's strive to become a global city (Wu 2000b). Unlike Guangdong's greenways, where the local leader understood the direction of the central government and built greenways to gain political recognition, Chengdu seemed to conduct the greenway project first as its own ambition for economic development but 'incidentally' received the recognition from the central endorsement. Following the endorsement, the city has escalated its green development. In this way, there has been a remarkable continuity of urban development strategy.

The park city is a vision to develop green infrastructure such as parks for the whole metropolis, arguing that rather than discrete parks within the city, now the city is situated in a vast region of parks. Urban built-up areas, greenways, and parks are interwoven and mixed. Beyond this abstract imagineering, it is a mixed-function development in the suburbs and even exurbs. As such, the greenway in Chengdu is moved more toward 'development' than the PRD greenways. It furthers President Xi Jinping's vision of 'green mountain, clean water' toward the urbanization of nature. Therefore, the Chengdu Greenway project was a project of the then-party secretary. He is the actual operator, although he did not claim his ownership and insisted that it is a project envisioned by the top leader. This is a political skill necessary in a multi-scalar state.

The Chengdu Greenway is more than 'greenwashing' or boasting green without developing green space. Compared with the PRD greenway, Chengdu invested more substantially in green space development. It also cleared up 'low-quality' development, i.e., the expansion of rural areas into the protected agricultural land. The result is that more green space claimed back from land encroachment, and some agricultural land might be converted into landscaping. It also has a narrow understanding of nature: the 'wild,' underdeveloped, and agricultural land are 'urbanized' through landscaping.

However, the effort of the party secretary might not be a great success for his political career. Fan Ruiping did not receive further promotion. While local cadres' promotion is used as an explanation for their entrepreneurial behaviors in the literature (see Chapter 2), in the cases of Wang Yang and Fan Ruiping, we do not know whether they intended to build the greenways for this specific purpose or whether the central government considers the greenway a success. The central government discovered the misuse of agricultural land from the Third National Land Survey and required Chengdu to restore it in mid-2021. The order was reinforced following President Xi's visit to Sichuan in June 2022 to review the agricultural land protection work. In the spring of 2023, restoring some green fields in greenways to return to agricultural use caught wide media attention (Figure 7.5).

After introducing two major policy initiatives and practices in eco-cities and greenways, we now turn to the two components of governance: the state and the market.

State strategies: ecological civilization

Ecological civilization

Although the concept of 'ecological civilization' was proposed before President Xi Jinping's era, more or less as a Chinese response to 'sustainable development' in the world, the term is now prominently attached to his environmental vision. It suggests a more fundamental

Figure 7.5 The greenways converted back into agricultural land in the 'Park City' of Chengdu.

understanding of the environment beyond its utilitarian value to the economy (Pan 2016; Goron 2018). It is more than a policy rhetoric, representing a state strategy—a new rationality—to solve the tension between economic growth and environmental protection. In 2018, ecological civilization was endorsed by the constitution of China.

This new rationality aims to achieve social and ecological sustainability as a continuing civilization—implying territorial politics, as when the concept is related to China, it is a Chinese civilization with a long history. Hence, maintaining a good environment, like economic development, is the pathway to civilization, as part of the 'Chinese dream'—reviving national prosperity, or China's modernization, as shown in Chapter 2 about a long-lasting effort of city planners originating from the Republic of China.

Therefore, stressing ecological civilization goes beyond the narrow focus on GDP growth. The state strategy aspired to shift from development to civilization. This means the policies should go beyond an economic-centered development principle—the pursuit of economic growth and profit maximization—and treat the environment as a priority in environmental governance. Further, ecological civilization emphasizes ecological efficiency and the quality of social reproduction (Pan 2016).

Originating from China, ecological civilization reflects new territorial politics, which aims to present an alternative to green capitalism or 'ecological modernization' (Goron 2018; Zhang and Wu 2023; Xie et al. 2023).[5] The concept of ecological modernization means the application of innovative, modernized, and experimental governance to facilitate eco-technological transition. The concept is more neutral in terms of governance. But conceived in

Western market economies, the concept means relying on the private sector to deliver environmental projects. Ecological modernization, in essence, is a market-oriented neoliberal governance in which the government may proactively engage with the private sector but still follows the route of modernization driven by economic growth.

In contrast, ecological civilization reflects more state-centric environmental governance, as civilization is not equivalent to modernization or economic development. As such, market-oriented development might be helpful to achieve this rationality, but it should not be the dominant consideration. The principle, thus, is not to maintain the environmental quality or fix its problem to sustain economic growth—a 'sustainability fix' as explained earlier in this chapter—but rather, it is the other way around: the state deploys any instrument possible, including market-oriented development, to realize its aspiration for civilization. This state-centered governance, as discussed in Chapter 1, is interpreted as 'state entrepreneurialism.'

The discourse of ecological civilization reveals the nature of state entrepreneurialism in environmental governance. The state utilizes market instruments to fulfill state objectives because ecological civilization is a state-proposed vision. The approach forefronts state strategic considerations. As will be seen later, such a vision practically leads to strengthened environmental policies and re-centralization toward the central government. As an 'art of governance' seen from a governmentality perspective, ecological civilization means that the multi-scalar state and society must follow this rising nation-state agenda.

In other words, ecological civilization leaves discretion to the governments at various levels, enterprises, and society to use their tactics to achieve this vision besides the top-down commands and policies. Ecological civilization is more than rhetoric. Instead, it is a political mandate—a form of environmental governmentality combining political rationalities and governmental technologies (Zhang 2021). While leaving some discretion to localities, it imposes an authority through the multi-scalar state. Its interpretation is not subject to an open debate or everyday life localities.

The multi-scalar state

To understand how ecological civilization, as a state strategy, affects China's environmental governance, we must interrogate the multi-scalar state. China is a large country with multiple levels of administration ranging from the central government and provincial government to the government of prefecture cities, cities, districts and countries, towns, and finally to the quasi-government agencies of street offices in the cities and townships (Wu 2002; Ma 2005; Li and Wu 2012; Lim 2019; Su 2022; Zhang et al. 2022). China's political geography reveals the importance of the administrative rank of cities. While the subnational states have their interests in localities and engage in inter-city competition, they are subject to upper-level government control.

Similarly, environmental governance is not simply a territorially conditioned endeavor. Economic devolution under economic reform, especially the greater local financial autonomy to raise capital from land development, has driven more growth-oriented urban governance in local economies. In environmental management, local governments did not fully implement the environmental policies promulgated by the central government, leaving an 'implementation gap' (Kostka and Nahm 2017). But at the same time, the central government strengthened its control over local governments through top-down monitoring measures such as the 'lifelong responsibility system,' which improved the effectiveness of implementing the central government's environmental policies (Kostka and Nahm 2017; Kostka and Zhang 2018).

In climate change governance, the central government demonstrates its enduring authority and top-down control (Westman et al. 2019). In land management, the central government allocates land development quota. It imposes stringent control in agricultural land protection and conservation planning, and the 'ecological enclosure' phenomenon is much more 'centrally coordinated, locally adaptive and self-interested' (Rodenbiker 2020, p. 696). As discussed in Chapter 2 on governance, the structure of the multi-scalar state means that local government has to absorb the objectives beyond the local territory. Local development is not driven by a growth machine formulated locally (Shen et al. 2020; Xie et al. 2020; Chung and Xu 2021; Zhang and Wu 2022b; Wu et al. 2022b).

The multi-scalar state suggests a more complex form of governance. In environmental governance, authoritarian environmentalism describes the authority of the central government. Fragmented authoritarianism indicates decentralized and conflicting government levels and departments. The multi-scalar state suggests a more nuanced understanding of the central–local relation. The central and local states have different motivations and priorities in environmental governance.

The central government has a stronger concern over the environmental crisis. However, local governments are often more concerned with the means to realize the environmental policy and have additional interests in local growth and revenue. These dynamics can be seen in eco-cities. The central government initiated the pilot program, namely the former Ministry of Housing and Urban and Rural Development. The program was intended for ecological preservation, green space creation, and low-carbon economies. Local states needed an endorsement from the central state for their eco-cities projects. This endorsement came with favorable support toward local development, such as more land development quotas (Chien 2013).

Therefore, local states must internalize the central state's policy objectives in constructing eco-cities. But at the same time, they need to mobilize financial resources—such as a land-financed development approach—at their discretion. This additional motivation drives them away from implementing the central government policy and seeking their interests. Often, they have to balance economic growth and environmental control. The rationality of ecological civilization imposes a new situation for the local states to perform the socio-ecological fix.

In the central–local relation, less attention has been paid to the provincial government. The provincial government does not directly invest in development projects as a local government. It acts as the mediator between the central government and municipal governments. Like the central government, it initiates environmental improvement projects under the general mandate of the central government. For example, Jiangsu province launched rural village improvement programs to provide necessary public hygiene and village infrastructure. It initiated a stringent campaign to control pollution and enforce energy transition. Like the central government, the provincial government enforces the national policies through key performance indicators to evaluate projects and cadre appointments. Compliance with environmental objectives is introduced into the compulsory performance indicator. [6] In Yunnan, the provincial government implements the central government policy by sending its provincial development corporation to construct the ecological corridor in Dali to protect Erhai Lake.

Facing diverse interests in the multi-scalar state and tensions between economic growth and the environment, ecological civilization imposes a new rationality or governmentality. It helps the central state achieve its goal toward more significant environmental concerns, not through concrete policies and direct commands. Instead, as a discourse, it hopes to motivate local governments and a wide range of non-state actors to achieve state strategies while leaving the actual pathway and implementation flexible and vague. Such governmentality includes the overall rationality of ecological civilization and governance technologies such as pilot programs, experiments, and performance indicators (Heilmann 2008; Zhang 2021).

In sum, ecological civilization is more than rhetoric or 'greenwashing.' It imposes a governmentality that triggers the power transformation across the multi-scalar state. It is similar to 'eco-state restructuring,' where the priority, such as climate change, generates state rescaling and restructuring.

Market operation: the development corporation

Although the state maintains its centrality in environmental governance, it deploys market instruments to operate environmental projects such as greenways. The state also creates its own agencies, such as urban development and investment corporations (UDICs), to finance and organize land development. The market actors also include private entrepreneurs in green industries. In some cases, like solar water heaters and the green economy, private entrepreneurs play a crucial role in the low-carbon transition (Yu and Gibbs 2018; Zhang et al. 2021). Real estate developers are major actors in the development of eco-cities. These eco-cities are mega-urban projects rather than government environmental programs. Even in a government-led program of greenways in the PRD, real estate developers were asked to contribute green space to the greenway projects. Thus, their profit-making motivation is accommodated in the overall project design (Chung et al. 2018).

Environmental governance presents a feature of entrepreneurialism in the green economy and eco-cities. As China adopted a market-oriented development approach, private entrepreneurs play an important role in the green economy. Like governance in innovation (see Chapter 5), where enterprises are the source of innovation, environmental governance relies on enterprises to deliver environmental projects.

How do market actors contribute to environmental governance and 'socio-technical transition'? As discussed in the theoretical section, the 'multi-level perspective' explains how green innovations emerged initially as the niche of technology and then upscaled into the landscape. This often occurs when the existing socio-technical regime experiences some external shock, forcing it to move away from the status quo. The role of private entrepreneurs is crucial in this process. For example, in Wuxi of Jiangsu province, the cyanobacteria bloom in Taihu Lake generated a landscape shock, pushing its energy transition toward solar industries. This allows the initial products in solar industries to become a major sector in Wuxi (Zhang et al. 2021). Key enterprises led this energy transition. The external shock turned the local state from preserving the existing energy regime to supporting green products.

While private entrepreneurs lead the green industries, the relationship between the private sector and the local state also affects the low-carbon transition. The socio-technical transition needs to be considered together with the multi-scalar state (Zhang et al. 2021). As shown in the city of Dezhou, where the solar water heater industries mainly relied on key private entrepreneurs, the relationship between the local government and private entrepreneurs was torturous. The government initially supported the industry in building a solar city. However, when the market changed, the government took action to restrict and restructure the industry (Yu and Gibbs 2018, 2023).

When we closely examine the market actors, we may find that the key actors include private entrepreneurs and state-owned enterprises. The vital and strategic industries are controlled and dominated by SOEs. In major environmental infrastructure projects, they operate in the market as government agencies. For example, in Chengdu, the development corporation under the municipality implemented the greenway project, which greatly sped up the greenway construction. Before the launching of greenways, district governments collaborated with private developers to develop green spaces in association with their real estate projects. This resulted in a quite scattered pattern of green spaces. In 2018, the municipal government launched a dedicated greenway development corporation under the municipal-controlled development corporation (*chengtou*) to carry out greenway development. This 'integrated operation' allowed the concentration of land resources and more effective land value capture. The development corporation borrowed from the capital market to finance the greenway project.

In Chengdu greenway development, the introduction of the municipal-level development corporation has strengthened the municipal government's authority because it does not have to negotiate with district governments for their financial contribution toward the greenway project. The corporation has opened up access to the financial market for the municipal government. This is a debt-driven approach because the greenway does not directly generate profits.

The development of the greenway is costly. The construction cost is planned to be covered by land sale in the process of greenway development and afterward in the future. The district governments of Chengdu bear the initial cost of land compensation with the hope of gaining income from land sales. The development corporation covers operational costs through income from the business and facilitates inside or along the greenway, including related hotels, restaurants, and entertainment facilities. Because of this funding requirement, the development corporation has to manage the business through development rather than containing development. It must consider the development costs and profitability because it is an agent operating in the market.

However, the state still influences the appointment of the board of directors in SOEs because of state ownership and institutional connection. The development corporation behaves differently from a shareholding company that often prioritizes profitability. The municipality created the greenway corporation with the mission of ecological civilization. The corporation has to consider the political pressure on the municipal government. In this way, the corporation's operation is subject to multiple objectives, and by establishing the corporation, the state exerts a vital effect on environmental governance (Feng et al. 2023).

Understanding the market operation is important to reveal the nature of environmental governance. The state does not directly command the environmental projects, as the development corporation has to negotiate with other market actors and even government bodies at different scales (for example, the municipal-level development corporation needs to collaborate with district governments). In this way, environmental policies are implemented through 'governance' rather than by the government. On the other hand, the development corporation's close association with the government shows that China's environmental governance no longer conforms to 'environmental neoliberalism.'

Further, from the governmentality perspective of ecological civilization, using market agents such as development corporations is a governance technique, in addition to other techniques such as rescaling central-local relations, 'environmental experiments,' and developing environmental discourses.[7] For example, China's low-carbon environmental governance includes various techniques such as compiling inventories, developing low-carbon plans, and promoting international collaborations (Zhang 2021, 2023). De-carbonization is a political rationality that follows the broad ecological civilization.

Society participation: marginalized but emerging through environmental activism

Thinking about the relationship between state and society, China is often regarded as an authoritarian nation. The role of society in environmental governance is marginalized. Market-oriented development creates greater social inequalities and negative impacts on vulnerable social groups. People with low incomes in underdeveloped rural areas bear a disproportional burden of the environmental cost. The backward economic condition forces them to engage in polluting and harmful industries (Wang et al. 2021, 2023b). They are victims of rural industrialization.

Moreover, low-income workers lack the economic and social resources to protest. Urban pollution also impacts the middle class in cities. They tend to be more aware of the problem and active in protest. The urban middle-class activism successfully transformed some environmental projects, such as the relocation of the PX project in Xiamen and the neighborhood zero-waste policy by the anti-incinerator movement in Guangzhou (Lin and Kao 2020). Society also bears the cost of environmental projects intended to save the environment. These projects generate unexpected or uneven consequences for some social groups, for example, relocation and displacement (Rodenbiker 2020).

In addition to citizen activists, environmental non-government organizations (ENGOs) appeared. Some have a background in government organizations or are sponsored by the government as government-organized NGOs (Shapiro 2016). They began to mediate between the state and society. In environmental governance, public participation is driven by the rising interests of the public. It became the most active in all forms of political participation because it was regarded as more technical, neutral, and less confrontational.

For a certain period, participation in environmental governance was more open to green civil society. In some areas, for example, waste management, NGOs are very active in environmental campaigns. However, they raise a profound question about the decision-making of development projects. Public participation thus becomes political. The government then tightened the regulation of green civil society. So far, environmental activism mainly resorts to the procedure and contests by the regulation and law. The recognition of environmental justice still needs to be improved or made more explicit in the agenda of environmental governance.

From the perspective of state entrepreneurialism, we can see that the state uses a variety of instruments, including government-made 'market' actors and public participation, to achieve its governance objectives, such as ecological civilization. Besides the promotion by the government, public participation also appears in some environmental issues through a bottom-up process. The state encourages public participation to mitigate conflicts and contestations, particularly at the grassroots. For example, community gardens are maintained by variegated approaches through resident involvement, external middle-class experiments, and state encouragement (Mai et al. 2023). Recently, in neighborhood regeneration, the state deliberately promoted public participation by appointing 'neighborhood planners' and organizing workshops and consultation meetings, as discussed in Chapter 5 about 'micro-redevelopment.'

In sum, examining the role of society provides a more nuanced understanding of environmental governance under state entrepreneurialism. When interrogating the governance dimension, we appreciate that so-called 'authoritarian environmentalism' (Eaton and Kostka, 2014) does not mean an absolute power concentration to the central government. There is a complex central–local relation, showing both the local discretion (Kostka and Nahm 2017) and the recent re-centralization of governance toward the central government under 'ecological civilization.'

Yet, looking at the state–society relation under ecological civilization, we also witness a plural of participation activities, rising environmental activism, and bottom-up neighborhood co-governance (Mai et al. 2023). Therefore, China's environmental governance has seen a rising environmental agenda—environmentalism. It also means a more state-centered approach—state entrepreneurialism—instead of neoliberalism as a state retreat. The state does not command directly, as seen in the governmentality. This also contradicts 'authoritarianism.'

The concept of state entrepreneurialism thus helps to provide a more nuanced understanding of Chinese environmental governance. It means state centrality in environmental governance, not a centralized governance structure.

Critical reflection

China's environmental crisis is a crisis of 'neoliberal' environmental governance after it introduced the market mechanism. The notion of neoliberal governance here does not refer to the absence of the state's role. Instead, it suggests that the state adopts a pro-market approach, especially attracting foreign investment to sustain economic growth and achieve national modernization. The state prioritized economic development in the earlier stage of economic reform in the 1980s and 1990s. The environmental quality was sacrificed for industrial growth. Pro-growth governance is not only a national strategy but also a policy adopted by the local governments, which were pressured by the central government and incentivized by economic institutions such as tax-sharing, which made them more reliant on income from land development.

China's entry into the World Trade Organization (WTO) symbolizes a greater integration of its economy with the global economy. However, the governance model of the 'world factory' requires a different role of the state. It deviated from the earlier neoliberal pro-market approach. While the state still strived to attract foreign investment, it began paying more attention to environmental quality. Because the world-factory model was a large-scale development, different from earlier village-based export-oriented material processing industries (Wang et al. 2023a), industrial districts were established, creating even more significant environmental impacts.

However, this 'world-factory' model is an urban model of social management, requiring the state to provide infrastructure and social services. The creation of development zones was achieved through 'planning for growth' (Wu, 2015b). Instead of dismissing the function of planning, it requires planning infrastructure. Providing comprehensive conditions is necessary for the production of the world factory. The environmental condition is among such conditions. Environmental governance is thus a 'socio-ecological fix' for capital accumulation, solving the problem of pollution and ecological degradation that prohibit sustained growth.

The global financial crisis in 2008 led to a 'financialized state entrepreneurialism' (Wu 2023a, 2023b). As discussed in Chapter 4 on financialization, local government financial vehicles (LGFVs) proliferate. Through LGFVs, the state gained development capital. This gave the local state more power to realize its development aspiration. As one major type of LGFVs, urban development and investment corporations (UDICs) take both development and financing functions. This means that the local state has its operational agents to take on concrete tasks of land development. At the same time, the deepening of socio-ecological crises seemed to challenge state legitimacy.

For the central state, this urgent issue gradually occupied the priority of the policy agenda, as seen in the shift toward 'ecological civilization.' As China has a multi-scalar state, the political mandate of the central government exerts a strong influence over local governments. They take the environmental mission and operate it at the local level. But at the same time, they are interested in environmental projects as these projects may also generate economic opportunities and benefits. The mandate of ecological civilization thus penetrates the multi-scalar state, and the socio-ecological fix is performed accordingly across urban and regional levels.

Because in the earlier stage of market reform, China's environmental governance shared some 'neoliberal' features, urban development was driven by the dynamics of the growth machine. Various projects under environmental initiatives, such as eco-cities, are deemed as rhetoric or greenwashing. On the other hand, the green turn under ecological civilization seems to suggest strengthened environmentalism, moving away from the dynamics of the growth machine. This raises the question of whether the growth machine or ecological civilization determines China's environmental governance. There is the tension between neoliberal and state-centered governmentalities—as the 'art of governance' using market or state agencies. The dichotomy, however, is only perceived rather than ontological (Zhang et al. 2022). As can be seen from environmental governance, state entrepreneurialism explains state centrality and market instruments as complementary rather than contradictory.

China's environmental governance demonstrates a complex governmentality. It is not by the growth machine for capital accumulation, nor is it a material translation of the political rhetoric—'ecological civilization.' The two co-exist and have different weights in policy-making, varying over time. Various levels of government have other priorities. When implementing the mandate of ecological civilization, local governments have to interpret and translate it into concrete tasks and projects. They must rely on the market mechanism and actors beyond the state and deal with vast economic interests. As explained earlier, as they work with and through the market, they give more practical considerations to the market operation. The local state is more oriented toward the growth machine dynamics. On the other hand, the central government considers more strategic challenges, especially the crisis brought about by 'neoliberal' governance and new geopolitical challenges. Because of the nature of the multi-scalar state, or 'fragmented authoritarianism' (Lieberthal and Lampton 2012), there has been an 'implementation gap' (Kostka and Nahm 2017).

As such, the objective of ecological civilization creates plural implementation approaches. Local governments sometimes strive to use environmental projects to promote economic growth—through place branding and city marketing, as explained by the 'entrepreneurial city.' On other occasions, they conduct environmental experiments through state-centered initiatives to demonstrate their alignment with the central government mandate. In contrast to market-oriented operations, they are more party-state endeavors. We witness variegated environmental practices and political ecologies in conservation plans, waste management, and low-carbon energy transition.[8]

Even at the local state level, China's environmental governance, seen from state entrepreneurialism, deviated from neoliberal governmentality. Although environmental policies resort to the market mechanism, the rationality of environmental governance is more state-centered. The multi-scalar state performs various socio-ecological fixes, hoping to solve the environmental challenges. The operation involves private-sector actors, state-owned enterprises, and state agencies.

Current environmental governance in China involves collaborations within the multi-scalar state and between the state and market agencies. The involvement of society has been limited. The degradation of environmental quality created negative impacts on people's livelihoods. The socio-ecological fix, intended to solve environmental problems, may actually create further impacts on society through displacement and green gentrification. The fix may also aim to solve economic and social issues. For example, in Chongming Island, Shanghai, the green requirement is deployed to promote economic restructuring. This green development is hoped to solve the social problems associated with the underdeveloped economy.

Similarly, Wuxi cleared up industrial uses along Taihu Lake to provide space for creative and new industries. The concept of the socio-ecological fix does not preclude the possibility of a greener outcome, while the previous studies on China's eco-cities regarded these projects primarily as greenwashing. Rather, it indicates that such a fix can harm different social groups. From the perspective of 'urban political ecology' (Keil 2020; Gandy 2022), the production of nature, in the form of the capitalist production of commodities, has social implications.

Recently, China has seen rising environmental activism and a 'green civil society.' People's everyday practices might change the course of environmental agendas or specific environmental projects. However, the state seems to dominate society on the environmental agenda, which is noted as 'authoritarian environmentalism' (Eaton and Kostka 2014). State entrepreneurialism describes the actually existing environmental governance with limited society in environmental decision-making. It does not preclude the possibility of a greater or 'self-governing' society in dealing with environmental issues in the future.

In addition to market instruments, the state strives to mobilize society to improve environmental quality and implement its sustainable agenda. The involvement of society, through either top-down campaigns or bottom-up middle-class initiatives, is welcomed as long as it supports a somewhat flexible governmentality of 'ecological civilization.' Society plays an increasingly significant role in everyday spaces such as 'community gardens' (Figure 7.6) and the compulsory program of neighborhood waste-sorting, which require more residents' collaboration. However, different social groups have unequal capacities to affect environmental projects and receive differentiated consideration in environmental governance under state

Figure 7.6 A community garden developed through the support of a professional organization in Shanghai Xuhui District.

entrepreneurialism, according to their relevance to the priority of state missions, which raises the issue of environmental justice.

Finally, this chapter shows that to understand China's environmental governance, we must combine a political economy understanding, especially the state–market–society relationship, and knowledge of ecological innovation and techniques. The MLP is a relatively generic and helpful perspective, which can be made more concrete in different cultural contexts (Huang et al. 2021). Its usefulness is a parsimonious description of a bottom-up process in response to external shocks, transforming its socio-technical regimes. Rather than provide a detailed account of these 'niches' and 'shocks,' it is helpful to understand the 'rationality' (here ecological civilization) embedded in environmental governance and the 'fix' associated with its political governmentality. We demonstrate that the socio-ecological fix in China reveals a more conscious state action, with its intentionality, to cope with a series of challenges, of which the environment metabolism is fundamental.

Notes

1 Its emissions reached over 10 Gt in 2020, accounting for almost 30 percent of global emissions. The country has been experiencing enormous pressure from the world. Nevertheless, please note that these figures do not consider the emissions generated by actual consumption. So, emissions from the production of goods made in China and consumed in the US are included in China's total emissions. In terms of CO_2 emission per capita, the Chinese average was 7.1 metric tons annually, lower than the US's 14.1 and higher than the average European countries. Suppose the numbers are adjusted for emissions from consumption rather than production. In that case, the per capita emission of China is reduced by 70 percent, to 5.4, but still above the world average of 4.8. For more information, please refer to: https://www.iea.org/reports/global-energy-review-co2-emissions-in-2021-2.

2 During the smog, millions of vehicles were forced off the road, and factories and construction sites shut down. Large cities like Beijing and Shanghai experienced worse smog.

3 See Zhang and Wu (2023) for a comprehensive collection of these practices.

4 This is from the media. Former Vice Minister of MOHURD, Qiu Baoxing, openly questioned the effect of some eco-cities.

5 However, Liu and Lo (2021, 2022) describe ecological civilization as ecological modernization.

6 This is a governance technique, as in environmental governmentality, see Zhang (2021).

7 For a complete discussion, see section on governmentality, experiments, and discourses in Zhang and Wu (2023).

8 See a wide range of studies in the *Handbook on China's Environmental Governance* (Zhang and Wu 2023).

Conclusion

State entrepreneurialism, as a perspective, sheds light on the dialectic relationship between the state and capital. Following this perspective, we can see that even though the Chinese state appears powerful, it is still deeply subject to capital accumulation. As a model, state entrepreneurialism represents state centrality in governing urban development. State capitalism assumes the totality of capital in a well-developed Western capitalist society. Such a totality is yet to be established in China, as the country has not been a capitalist heartland or an 'epic center of neoliberalism' (Harvey 2005). Rather, it is a frontier where tension between capital accumulation and territorial politics is visibly displayed. The tension has given rise to a modernizing state with Chinese characteristics (Wu 2022a). Entrepreneurialism as a new governance technique—using market instruments, adapting institutional flexibility, and co-opting the society—greatly enhances state resilience.

To understand China's urban governance, this book starts by describing concrete practices and processes governing urban development. China started with a low level of urbanization. Only about 19 percent of its population was classified as urban, meaning they were entitled to state welfare, just before economic reform started in 1979. The level of urbanization has rapidly increased to 66 percent in 2023. China has become a majority urban country. In this period, China also experienced industrialization and joined in the process of globalization to become the world factory. With increasing personal wealth and consumerism, however, it has not seen a rising civil society. It does not see a particular form of neoliberal governance. Neoliberalism, as a perspective of the state–capital relation, offers an insightful observation of a particular response to the crisis of capital accumulation—through deregulation, greater capital mobility, and flexible accumulation. This is mainly because of the declining profit rate of Western capitalism, which is a result of the welfare state. Neoliberalism is impossible in China since the state itself is a market builder. Chinese urban studies confirm that a self-regulating market is only an illusion—an insightful conclusion derived from the Western historical social transformation (Polanyi 1944).

As the economic reform introduced the market mechanism, marketization, accompanied by urbanization, dismantled the social foundation of rural China. In the period of state socialism, some features of collectivism of the traditional society were maintained at least partially (Wu 2022a), as a modern citizenship-based welfare state was underdeveloped. Less than 20 percent of the population was entitled to state welfare, mainly provided by workplaces. Even though the market has existed for a long time in China, its rationality has been controlled. The reform has introduced market operation. However, market rationality does not replace state rationality as the dominant ideology of governance. The state—as a formal set of social relations—and its apparatus—as a set of institutionalized practices—have not

DOI: 10.4324/9781003355175-9

been converted into capitalism. But, similar to neoliberalism as a class restoration project, state entrepreneurialism is also a state project, representing an existing class to use 'flexible' accumulation and governance experiments. Urbanization represents such a flexible form—expanding the space of capital accumulation globally, as Chinese rural migrants work for the world factory. As China has not seen the full-fledged formation of a capitalist class, class dominance does not come from the capital alone; it is reinforced through territorial politics.

Even though this book highlights territorial politics, it does not attribute the logic of governance to the bureaucratic organizations of the state, which can be regarded as the formal component of the state. In other words, although state entrepreneurialism identifies state centrality as a model in China, it is not a state-centered perspective. It is a political economy perspective of urban processes. The starting point is not the state. State entrepreneurialism is an explanation of emergent urban governance. Like other variegated neoliberal governance in advanced capitalist economies, the notion of 'governance' needs to consider the political economy of state–market–society relationships. The logic of governance is beyond the state itself or the state logic.

State entrepreneurialism represents such logic in governing capital circulation throughout the whole process. Such a view of totality means that China's urban governance is comparable to other concrete forms such as urban entrepreneurialism (Harvey 1989), austerity urbanism (Peck 2012), or municipal entrepreneurialism (Lauermann 2018). Urban entrepreneurialism, as a perspective, indicates the rising role of capital in late capitalism. It widely applies to diverse contexts in established and emerging market economies, including China. As a specific model of local politics reacting toward capital mobility, inter-city competition, and place marketing, it has been challenged as time changes in advanced market economies.

Austerity urbanism indicates that such reactionary entrepreneurialism becomes meaningless as the state is subject to a financialized governance logic—imposed by financial capital (Peck 2012). It is the deepening of neoliberalism, but public entrepreneurialism ended. The state has hardly any resources to promote a progressive agenda. On the other hand, municipal statecraft depicts a contrasting picture of progressive social movement at the municipal level. Outside the neoliberalism heartland, the specific form of urban governance is even more diverse. The rising influence of capital is observed in a traditionally strong society with greater informality or a strong state with a tradition of developmentalism. These economies have been involved in a global economy to varying degrees, through either the flexible accumulation of advanced capitalist economies or their national 'worldling' strategies in the Global South (Roy and Ong 2011). These latter practices are similar to China's market development strategies, involving different territorial politics other than the growth machine.

Following Harvey's urban entrepreneurialism perspective, China has seen a rising de facto capital influence, an insight from neoliberalism. On this front, neoliberalism, as a perspective, is relevant to China. However, China is not a modified model (with Chinese characteristics) of his specific reactionary neoliberal state, derived mainly from the US case. China's urban development model is export-oriented, land-driven, and financialized. Its governance is characterized by a salient state role and instrumental use of market tools (Wu 2018b). Its ideology is away from neoliberalism. Instead, the Chinese dream is presented as 'common prosperity'—an ideal rather than an outcome. The contribution of state entrepreneurialism to the critical urban studies literature is that it stresses the role of the state (Wu and Zhang 2022).

Harvey's neoliberalism highlights the role of capital. His theory is often read as a deterministic capitalist logic. As mentioned in the Introduction and Chapter 1, his characterization of neoliberalism as a state project of class restoration and using the terms of urban entrepreneurialism and flexible accumulation interchangeably allude to state actions, suggesting that his political economy explanation is less deterministic than we might think. His term 'accumulation by dispossession' is not deterministic but rather frontloads the capitalist state and territorial politics and strategies. In all these conceptualizations, he insists on seeing the capital movement in its totality, within which the state is part of capital relations (Harvey 2023). On this front, the perspective of state entrepreneurialism shares the same theoretical ground with neoliberalism. Both reflect state actions, but the concrete approaches—statecraft— differ. State entrepreneurialism helps reveal this implied totality more explicitly—as state sovereignty and territorial politics are involved in capital accumulation.

As a non-capitalist society, the market existed before China was subject to global capitalism. But its rationality is constrained and subject to the Confucian ideology, and the principle of societal organization is the 'differential mode of association' (*chaxugeju*) in a society known as 'rural China' (*xiangtu zhongguo*), as a theoretical construct, unlike the urbanized city-state with citizenships (Fei 1947/1992). China is a place to observe territorial politics during urban transformation. As China is becoming urban, the relationship between capital accumulation and territorial politics is subject to dialectic changes. Contrary to the view of some critics that the attention to the state may lead to China's exceptionalism, state entrepreneurialism makes the Chinese case even more comparable to similar dynamics in which globalization and geopolitics are coupled in post-pandemic state capitalism.

This book demonstrates the usefulness of critical urban studies. It is a political economy study of China's urban development, highlighting the introduction of market mechanisms and operations. Within this context, 'governance' emerges, in a form similar to what has been extensively observed in late capitalism. However, the Chinese experience reveals that the exact meaning of governance is not a power redistribution—the notion of neoliberal 'governing at a distance' is not applicable. Instead, the market and society are all instruments deployed by the state. It is in this sense that we see the continuing logic of governance. That is, economic reform does not fundamentally change the logic. It is characterized as market utilitarianism.

After a dialogue with critical studies on governance in Chapter 1, the historical formation of state entrepreneurialism and its definition features discussed in Chapter 2 stress the continuation of governance logic. Because of this continuation, the statecraft—exact governance form—differs from neoliberal governance, a more societal movement, and its form of 'municipalism' (Lauermann 2018). Statecraft originally refers to the art of staying in office, often assumed in a democratic political system, and maintaining the capacity to control. More broadly, it is an art of governance or the organizational dynamics of a bureaucratic system. The statecraft does not need to take an entrepreneurial form. Compared with urban entrepreneurialism, the statecraft seeks an explanation of managerialism. As mentioned in Chapter 4, financialization can be regarded as an internal managerial response to fiscal austerity rather than an entrepreneurial—pro-business—action. It would not be surprising to see state entrepreneurialism as a state-initiated managerial change. China has a long history of territorial management, planning, and resource mobilization, as discussed in Chapters 2, 3, and 4.

Beyond contextual specificities, the book provides an assessable narrative to explain the dynamics of city planning, urban development finance, urban redevelopment, local economic growth and innovation, and environmental governance. They are represented as versatile entrepreneurial statecraft (Wu et al. 2024). Although statecraft tends to use entrepreneurial means—either supporting market development or reforming public management through mimicking market operation—internalizing the market inside the state—the concept reflects more managerialism than entrepreneurialism. This is reflected in the intentionality of the state. That is, the statecraft combines capital accumulation and territorial politics. Statecraft is beyond capital accumulation and has multiple extra-economic objectives. These extra-economic objectives distinguish state entrepreneurialism from urban entrepreneurialism, as the former uses economic means to achieve its political intention—in China's case, the legitimacy and stability of the party-state, while the latter often regards a reflection of business interests. Although China seems extreme, state entrepreneurialism is more general than a particular pro-business form of governance, as contested by many recent studies on the variegated form of statecraft (Wu et al. 2024).

China's city planning for growth demonstrates the active role of the state in governing its built environment toward economic development. City planning has seen compounded and multiple objectives, ranging from expanding local government land-based finance to fulfilling various central government policies. The post-political notion suggests the market or capital dominance in the decision-making process. Within the critical urban studies literature, there is debate over whether post-politics reflects a tendency rather than an achieved condition. This tendency exists in China as planning becomes more technical, with market-based instruments and techniques. China's planning has never been properly political—through a Western democratic process. Post-reform planning reflects the centrality of political mandates not formulated and selected through a political process. If the post-political implications are taken this way, planning in China might be regarded as post-political, but it is due to a different governmentality from market-dominated rationality. It bears some similarity if post-political is read as a tendency away from the political debate. Planning in China is for the state, and its agenda is assigned through political mandate. The state intention is achieved through more professional planning techniques and public participation in planning processes to fulfill a defined agenda.

Like strategic guidance through city planning, resource mobilization has seen a new tendency toward a more financialized one in post-reform China. The 'property state' notion stresses the capitalist state as the landowner has a vested interest in the land and treats the city as a growth machine. Similarly, the Chinese state is deeply involved in land development and interested in land revenue. However, land-driven urban development is more than seeking land profits. The state uses land to mobilize capital in order to finance urban development. This capital mobilization involves various financial approaches, leading to financialization. Hence, in China, urban financialization does not originate from the finance sector, seeking a haven of capital deposits in the built environment. The state in the advanced market economies facilitates or enables financialization through deregulation and transferring more power to financial capital. In China, it is state-initiated. The state utilizes a financial approach to raise capital for development or other purposes, including refinancing the debt and dealing with the impact of financial crises.

After examining planning and finance, this book illustrates multiple objectives in urban redevelopment, innovation, and the environment. Chapters 5, 6, and 7 describe, respectively, the state's strategic intentions in urban redevelopment, innovation, and the environment. These intentions can be read as merely rhetoric, but they contain important messages about an ideological shift from 'growth-first' pragmatism. They convey a market critique rather than depicting a picture of market efficiency. Urban redevelopment emphasizes 'co-governance' to 'build the Peoples' City.' Innovation policies stress the aim for 'national indigenous innovation' and high-quality development. Environmental governance is mandated by 'ecological civilization.' Muti-level governments deploy these strategies to justify their local actions in addition to market signals. They all indicate that governing urban development does not follow a single objective of capital accumulation. These discourses reflect the territorial politics of the state.

This book has illustrated some distinctive features of China's urban governance, especially the role of the state in governing urban development. This salient role can be described as a *model* of state entrepreneurialism, reflecting the specificity of the Chinese case. Nevertheless, the role of the state is more universal across different places. Thinking about the relationship between the state and capital, state entrepreneurialism offers a *perspective* of their dynamics. The book has demonstrated how this perspective can be used to explain city planning, financing urban development, urban regeneration, local development and innovation, and environmental governance in China. It demonstrates that the neoliberal form of governance is a particular mode. In advanced capitalism, there are different forms, such as austerity statecraft or entrepreneurial municipalism. They all reflect the deepening of neoliberalism and post-political tendencies—a strengthened constraint of capital accumulation.

The Chinese case does not indicate a powerful and omnicompetent state. Rather, it reflects a severe constraint imposed on the state in the post-pandemic world and the state's response to the crisis of capital accumulation. Chinese statecraft as state entrepreneurialism represents a particular way of capital accumulation and related territorial politics (Wu et al. 2024). It aims to maintain state extra-economic intention and more strategic—meaning more top-down and *ad hoc* rather than well-planned and bottom-up—actions. However, the implementation of these intentions and the coordination of these actions resort to market and society. Often, the state creates its own agents in the market or co-opt those social actors emerging outside the state.

Entrepreneurial statecraft suggests that the state is not an entrepreneur, and its governance does not convert into a market-centered one. Rather, it is the state utilizing the market to serve its purposes. As this book shows, government officials often seek their career advancement or personal benefits—corruption—rather than profit maximization and market efficiency. Their investment decisions may not result in capital accumulation and expansion. Even if the land revenue is part of the consideration, it is mainly a means and a resource for fulfilling their objectives. More often than not, government officials are not competent in market operations. They ignore market signals and replace them with their wishes. They encouraged development corporations to borrow recklessly, resulting in low efficiency in resource utilization and redundant or unaccomplished projects. They behave differently from entrepreneurs because they do not follow market rationality.

State entrepreneurialism helps reveal the tension between the state and capital rather than a smooth coupling process. Compared with state capitalism, the tension between capital accumulation and territorial politics is more visible under state entrepreneurialism. Qing dynasty reformers introduced Western technologies but rejected capitalist institutions. Deng Xiaoping's pragmatism believed the market could also exist in socialism if it were an efficient technique. The reform did not completely embrace market or capitalist logic. The tendency is to reinforce territorial politics: the 'Chinese style of modernization,' the Chinese Communist Party (CCP) centrality, and upward accountability of state apparatus. When the state uses land to raise capital for development and deal with financial risks along with land-driven development, it does not follow the financial logic; rather, it is a developmental logic. Chinese examples expand the scope of inquiry and enrich these concepts of critical urban studies, presenting a fuller picture of governing urban development in the 21st century.

References

Aalbers, M. B. (2008). The financialization of home and the mortgage market crisis. *Competition and Change*, *12*, 148–166.

Aalbers, M. B. (2016). *The Financialization of Housing: A Political Economy Approach*. London: Routledge.

Aalbers, M. B. (2017). Corporate financialization. In D. Richardson, N. Castree, M. F. Goodchild, A. Kobayashi, W. Liu, & R. A. Marston. (Eds.), *International Encyclopedia of Geography: People, the Earth, Environment and Technology* (pp. 1–11). Oxford: Wiley-Blackwell.

Aalbers, M. B. (2020). Financial geography III: The financialization of the city. *Progress in Human Geography*, *44*, 595–607.

Aalbers, M. B. (2023). State/finance symbiosis. *Urban Geography*, *44*, 796–802.

Aalbers, M. B., Taylor, Z. J., Klinge, T. J., & Fernandez, R. (2023). In real estate investment we trust: State de-risking and the ownership of listed US and German residential real estate investment trusts. *Economic Geography*, *99*, 312–335.

Alami, I. (2023). Ten theses on the new state capitalism and its futures. *Environment and Planning A: Economy and Space*, *55*, 764–769.

Alami, I., Babic, M., Dixon, A. D., & Liu, I. T. (2022). Special issue introduction: What is the new state capitalism? *Contemporary Politics*, *28*, 245–263.

Alami, I., & Dixon, A. D. (2020). State capitalism (s) redux? Theories, tensions, controversies. *Competition & Change*, *24*, 70–94.

Alami, I., Dixon, A. D., & Mawdsley, E. (2021). State capitalism and the new global D/development regime. *Antipode*, *53*, 1294–1318.

Allen, J., & Cochrane, A. (2007). Beyond the territorial fix: Regional assemblages, politics and power. *Regional Studies*, *41*, 1161–1175.

Allen, J., & Cochrane, A. (2010). Assemblages of state power: Topological shifts in the organization of government and politics. *Antipode*, *42*, 1071–1089.

Allmendinger, P. (2016). *Neoliberal Spatial Governance*. London: Routledge.

Allmendinger, P., & Haughton, G. (2012). Post-political spatial planning in England: A crisis of consensus? *Transactions of the Institute of British Geographers*, *37*, 89–103.

Allmendinger, P., Haughton, G., & Shepherd, E. (2016). Where is planning to be found? Material practices and the multiple spaces of planning. *Environment and Planning C-Government and Policy*, *34*, 38–51.

Andrusz, G. M., Harloe, M., & Szelenyi, I. (1996). *Cities After Socialism: Urban and Regional Change and Conflict in Post-socialist Societies*. Oxford: Blackwell.

Appelbaum, R. P., Cao, C., Han, X., Parker, R., & Simon, D. (2018). *Innovation in China: Challenging the global science and technology system*. Cambridge: Polity.

Asheim, B. T., Boschma, R., & Cooke, P. (2011). Constructing regional advantage: Platform policies based on related variety and differentiated knowledge bases. *Regional Studies*, *45*, 893–904.

Ashton, P., Doussard, M., & Weber, R. (2016). Reconstituting the state: City powers and exposures in Chicago's infrastructure leases. *Urban Studies*, *53*, 1384–1400.

August, M., & Walks, A. (2017). Urban redevelopment. In D. Richardson, N. Castree, M. F. Goodchild, A. Kobayashi, W. Liu, & R. A. Marston (Eds.), *The International Encyclopedia of Geography*. London: John Wiley & Sons.

Bai, C.-E., Hsieh, C.-T., & Song, Z. M. (2016). The long shadow of a fiscal expansion. *National Bureau of Economic Research, Paper No. 22801*, 1–45.
Ban, C., & Bohle, D. (2021). Definancialization, financial repression and policy continuity in East-Central Europe. *Review of International Political Economy*, *28*, 874–897.
Bathelt, H., Malmberg, A., & Maskell, P. (2004). Clusters and knowledge: Local buzz, global pipelines and the process of knowledge creation. *Progress in Human Geography*, *28*, 31–56.
Beeson, M. (2010). The coming of environmental authoritarianism. *Environmental Politics*, *19*, 276–294.
Belotti, E., & Arbaci, S. (2021). From right to good, and to asset: The state-led financialisation of the social rented housing in Italy. *Environment and Planning C: Politics and Space*, *39*, 414–433.
Benjamin, S. (2008). Occupancy urbanism: Radicalizing politics and economy beyond policy and programs. *International Journal of Urban and Regional Research*, *32*, 719–729.
Beswick, J., & Penny, J. (2018). Demolishing the present to sell off the future? The emergence of 'financialized municipal entrepreneurialism'in London. *International Journal of Urban and Regional Research*, *42*, 612–632.
Beveridge, R., & Featherstone, D. (2021). Introduction: Anti-politics, austerity and spaces of politicisation. *Environment and Planning C: Politics and Space*, *39*, 437–450.
Beveridge, R., & Koch, P. (2017). The post-political trap? Reflections on politics, agency and the city. *Urban Studies*, *54*, 31–43.
Boschma, R. (2005). Proximity and innovation: A critical assessment. *Regional Studies*, *39*, 61–74.
Boschma, R., Heimeriks, G., & Balland, P.-A. (2014). Scientific knowledge dynamics and relatedness in biotech cities. *Research Policy*, *43*, 107–114.
Boschma, R. A., & Frenken, K. (2006). Why is economic geography not an evolutionary science? Towards an evolutionary economic geography. *Journal of Economic Geography*, *6*, 273–302.
Bremmer, I. (2010). *The End of the Free Market: Who Wins the War Between States and Corporations?* New York: Portfolio.
Brenner, N. (2004). Urban governance and the production of new state spaces in western Europe, 1960–2000. *Review of International Political Economy*, *13*, 447–488.
Brenner, N. (2019). *New Urban Spaces: Urban Theory and the Scale Guestion*. Oxford: Oxford University Press.
Brenner, N., & Schmid, C. (2011). Planetary urbanization. In M. Gandy (Ed.), *Urban Constellations* (pp. 10–13). Berlin: Jovis.
Brenner, N., & Theodore, N. (2002). Cities and the geographies of 'actually existing neoliberalism'. *Antipode*, *34*, 349–378.
Brownill, S., & O'Hara, G. (2015). From planning to opportunism? Re-examining the creation of the London Docklands Development Corporation. *Planning Perspectives*, *30*, 537–570.
Buckingham, W. (2017). Uncorking the neoliberal bottle: Neoliberal critique and urban change in China. *Eurasian Geography and Economics*, *58*, 297–315.
Bulkeley, H., Broto, V. C., & Maassen, A. (2014). Low-carbon transitions and the reconfiguration of urban infrastructure. *Urban Studies*, *51*, 1471–1486.
Bulkeley, H., Castan Broto, V., Hodson, M., & Marvin, S. (2010). *Cities and Low Carbon Transitions* (Vol. 35). London: Routledge.
Caprotti, F., Springer, C., & Harmer, N. (2015). 'Eco' for whom? Envisioning eco-urbanism in the Sino-Singapore Tianjin Eco-city, China. *International Journal of Urban and Regional Research*, *39*, 495–517.
Carmona, M. (2009). The Isle of Dogs: Four development waves, five planning models, twelve plans, thirty-five years, and a renaissance… of sorts. *Progress in Planning*, *71*, 87–151.
Castree, N., & Christophers, B. (2015). Banking spatially on the future: Capital switching, infrastructure, and the ecological fix. *Annals of the Association of American Geographers*, *105*, 378–386.
Caves, R.W. and Cullingworth, B. (2024) *Planning in the USA: policies, issues, and processes*. London: Routledge.
Chan, K. W. (1994). *Cities with Invisible Walls*. Hong Kong: Oxford University Press.
Chan, K. W. (2010). The global financial crisis and migrant workers in China: 'there is no future as a labourer; returning to the village has no meaning'. *International Journal of Urban and Regional Research*, *34*, 659–677.
Chang, I.-C. C. (2023). Eco-cities in China: National initiatives, local implementation and livelihood transitions. In F. Zhang & F. Wu (Eds.), *Handbook on China's Urban Environmental Governance* (pp. 227–242). Cheltenham, UK: Edward Elgar.

Chang, I.-C. C., Leitner, H., & Sheppard, E. (2016). A green leap forward? Eco-state restructuring and the Tianjin–Binhai eco-city model. *Regional Studies*, *50*, 929–943.

Chang, I.-C. C., & Sheppard, E. (2013). China's eco-cities as variegated1 urban sustainability: Dongtan eco-city and Chongming eco-island. *Journal of Urban Technology*, *20*, 57–75.

Chatterjee, I. (2009). Social conflict and the neoliberal city: A case of hindu-muslim violence in India. *Transactions of the Institute of British Geographers*, *34*, 143–160.

Chatterjee, P. (2004). *The Politics of the Governed: Popular Politics in Most of the World*. New York: Columbia University Press.

Chen, Y.-L., & Shin, H. B. (2019). *Neoliberal Urbanism, Contested Cities and Housing in Asia*. New York: Palgrave Macmillan. Springer.

Chen, Z., He, Z., & Liu, C. (2020). The financing of local government in China: Stimulus loan wanes and shadow banking waxes. *Journal of Financial Economics*, *137*, 42–71.

Chien, S. S. (2013). Chinese eco-cities: A perspective of land-speculation-oriented local entrepreneurialism. *China Information*, *27*, 173–196.

Christophers, B. (2010). On voodoo economics: Theorising relations of property, value and contemporary capitalism. *Transactions of the Institute of British Geographers*, *35*, 94–108.

Christophers, B. (2015). The limits to financialization. *Dialogues in Human Geography*, *5*, 183–200.

Christophers, B. (2017). The state and financialization of public land in the United Kingdom. *Antipode*, *49*, 62–85.

Christophers, B. (2019). Putting financialisation in its financial context: Transformations in local government-led urban development in post-financial crisis England. *Transactions of the Institute of British Geographers*, *44*, 571–586.

Chung, C. K. L., & Xu, J. (2021). Scalar politics of urban sustainability: Governing the Chinese city in the era of ecological civilisation. *Transactions of the Institute of British Geographers*, *46*, 689–703.

Chung, C. K. L., Zhang, F., & Wu, F. (2018). Negotiating green space with landed interests: The urban political ecology of greenway in the Pearl River Delta, China. *Antipode*, *50*, 891–909.

Cochrane, A. (2007). *Understanding Urban Policy: A Critical Approach*. Oxford: Blackwell Publishing.

Coe, N. M., & Yeung, H. W.-C. (2015). *Global Production Networks: Theorizing Economic Development in an Interconnected World*. Oxford: Oxford University Press.

Coe, N. M., & Yeung, H. W.-C. (2019). Global production networks: Mapping recent conceptual developments. *Journal of Economic Geography*, *19*, 775–801.

Cooke, P. (1992). Regional innovation systems: Competitive regulation in the new Europe. *Geoforum*, *23*, 365–382.

Cooke, P. (2004). The molecular biology revolution and the rise of bioscience megacentres in North America and Europe. *Environment and Planning C: Government and Policy*, *22*, 161–177.

Cooke, P., De Laurentis, C., Tödtling, F., & Trippl, M. (2007). *Regional Knowledge Economies: Markets, Clusters and Innovation*. Cheltenham, UK: Edward Elgar Publishing.

Cooke, P., & Morgan, K. (1998). *The Associational Economy*. Oxford: Oxford University Press.

Crouch, C. (2009). Privatised Keynesianism: An unacknowledged policy regime. *The British Journal of Politics and International Relations*, *11*, 382–399.

Cugurullo, F. (2021). *Frankenstein Urbanism: Eco, Smart and Autonomous Cities, Artificial Intelligence and the End of the City*. London: Routledge.

Dagdeviren, H., & Karwowski, E. (2022). Impasse or mutation? Austerity and (de) financialisation of local governments in Britain. *Journal of Economic Geography*, *22*, 685–707.

Davidson, M. (2019). Postpolitical city. In A. M. Orum (Ed.), *The Wiley Blackwell Encyclopedia of Urban and Regional Studies* (pp. 1–4). Oxford: Wiley.

Davidson, M., & Iveson, K. (2015). Recovering the politics of the city: From the 'post-political city' to a 'method of equality' for critical urban geography. *Progress in Human Geography*, *39*, 543–559.

Deng, H. (2023). Positioning China's state entrepreneurialism in structural coherence and multiple logics. *Transactions in Planning and Urban Research*, *2*, 296–313.

Dikeç, M., & Swyngedouw, E. (2017). Theorizing the politicizing city. *International Journal of Urban and Regional Research*, *41*, 1–18.

Doshi, S. (2019). The redevelopmental state: Governing surplus life and land in the 'urban age'. *Development and Change*, *50*, 679–706.

Doucette, J., & Park, B.-G. (2018). Urban developmentalism in East Asia: Geopolitical economies, spaces of exception, and networks of expertise. *Critical Sociology*, *44*, 395–403.

Du, Y. (2022). Holistic gentrification and the recreation of rurality: The new wave of rural gentrification in Chengdu, China. *Journal of Rural Studies*, *96*, 237–245.

Duckett, J. (2001). Bureaucrats in business, Chinese-style: The lessons of market reform and state entrepreneurialism in the People's Republic of China. *World Development*, *29*, 23–37.

Eaton, S., & Kostka, G. (2014). Authoritarian environmentalism undermined? Local leaders' time horizons and environmental policy implementation in China. *The China Quarterly*, *218*, 359–380.

Ekers, M., Hamel, P., & Keil, R. (2012). Governing Suburbia: Modalities and Mechanisms of Suburban Governance. *Regional Studies*, *46*, 405–422.

Ekers, M., & Prudham, S. (2017). The metabolism of socioecological fixes: Capital switching, spatial fixes, and the production of nature. *Annals of the American Association of Geographers*, *107*, 1370–1388.

Epstein, G. A. (2005). *Financialization and the World Economy*. Cheltenham, UK: Edward Elgar Publishing.

Etherington, D., & Jones, M. (2018). Re-stating the post-political: Depoliticization, social inequalities, and city-region growth. *Environment and Planning A: Economy and Space*, *50*, 51–72.

Etzkowitz, H., & Leydesdorff, L. (2000). The dynamics of innovation: From National Systems and 'Model 2' to a Triple Helix of university–industry–government relations. *Research Policy*, *29*, 109–123.

Evans, G. (2003). Hard-branding the cultural city–from Prado to Prada. *International Journal of Urban and Regional Research*, *27*, 417–440.

Evans, G. (2009). Creative cities, creative spaces and urban policy. *Urban Studies*, *46*, 1003–1040.

Fainstein, S. S. (2001). *The City Builders: Property Development in New York and London, 1980-2000* (2nd ed.). Lawrence, KS: The University Press of Kansas.

Fainstein, S. S. (2008). Mega-projects in New York, London and Amsterdam. *International Journal of Urban and Regional Research*, *32*, 768–785.

Fearn, G., & Davoudi, S. (2022). From post-political to authoritarian planning in England: a crisis of legitimacy. *Transactions of the Institute of British Geographers*, *47*, 347–362.

Fei, X. (1947/1992) *From the soil: the foundations of Chinese society (a translation of Fei Xiaotong's Xiangtu Zhongguo, by Gary G Hamilton and Wang Zhen)*. Berkeley, CA: California University Press.

Feng, Y., Wu, F., & Zhang, F. (2022a). Changing roles of the state in the financialization of urban development through chengtou in China. *Regional Studies*, *56*, 1259–1270.

Feng, Y., Wu, F., & Zhang, F. (2022b). The development of local government financial vehicles in China: A case study of Jiaxing Chengtou. *Land Use Policy*, *112*, 104793.

Feng, Y., Wu, F., & Zhang, F. (2023). Shanghai municipal investment corporation: Extending government power through financialization under state entrepreneurialism. *Environment and Planning C: Politics and Space*, *41*, 20–36.

Ferm, J., & Raco, M. (2024). Beyond entrepreneurialism: Revealing multiple logics of suburban housing development politics in the global city. *Territory, Politics, Governance*, 1–21. https://doi.org/10.1080/21622671.2024.2303333

Fernandez, R., & Aalbers, M. B. (2016). Financialization and housing: Between globalization and varieties of capitalism. *Competition & Change*, *20*, 71–88.

Fields, D. (2018). Constructing a new asset class: Property-led financial accumulation after the crisis. *Economic Geography*, *94*, 118–140.

Fields, D., & Uffer, S. (2016). The financialisation of rental housing: A comparative analysis of New York City and Berlin. *Urban Studies*, *53*, 1486–1502.

Florida, R. (2002). *The Rise of the Creative Class*. New York: Basic Books.

Florida, R., Adler, P., & Mellander, C. (2017). The city as innovation machine. *Regional Studies*, *51*, 86–96.

Follmann, A., Kennedy, L., Pfeffer, K., & Wu, F. (2023). Peri-urban transformation in the Global South: A comparative socio-spatial analytics approach. *Regional Studies*, *57*, 447–461.

Forrest, R., & Hirayama, Y. (2015). The financialisation of the social project: Embedded liberalism, neoliberalism and home ownership. *Urban Studies*, *52*, 233–244.

Forrest, R., Ren, J., & Wissink, B. (2019). *The City in China: New Perspectives on Contemporary Urbanism*. Bristol: Policy Press.

Foucault, M. 2008. The birth of biopolitics. New York: Palgrave Macmillan.

Foye, C. (2022). Section 106, viability, and the depoliticization of English land value capture policy. *International Journal of Urban and Regional Research*, *46*(2), 269–286.

Friedmann, J. (2005). *China's Urban Transition*. Minneapolis: University of Minnesota Press.

Fu, W., & Lim, K. F. (2022). The constitutive role of state structures in strategic coupling: On the formation and evolution of Sino-German production networks in Jieyang, China. *Economic Geography*, *98*, 25–48.

Fu, X., McKern, B., & Chen, J. (2021). *The Oxford Handbook of China Innovation.* Oxford: Oxford University Press.

Fu, Y., & Zhang, X. (2023). When an entrepreneurial government hammers out a plan for sustainable growth: A sustainable urban experiment story in China. In F. Zhang & F. Wu (Eds.), *Handbook on China's Urban Environmental Governance* (pp. 54–68). Cheltenham, UK: Edward Elgar.

Gandy, M. (2022). Urban political ecology: A critical reconfiguration. *Progress in Human Geography, 46*, 21–43.

Garreau, J. (1991). *Edge city: Life on the New Frontier*. New York: Doubleday.

Geels, F. W. (2002). Technological transitions as evolutionary reconfiguration processes: A multi-level perspective and a case-study. *Research Policy, 31*, 1257–1274.

Geels, F. W., Tyfield, D., & Urry, J. (2014). Regime resistance against low-carbon transitions: Introducing politics and power into the multi-level perspective. *Theory, Culture & Society, 31*, 21–40.

Geng, Y., Jiang, W., & Wang, Z. (2023). Planning centrality, state-oriented growth, and the spatial evolution of development zones in urban China: The case of Wuhan's Optics Valley. *Transactions in Planning and Urban Research, 2*, 115–131.

Gerlofs, B. A. (2019). Policing perception: Postpolitics and the elusive everyday. *Urban Geography, 40*, 378–386.

Ghertner, D. A. (2014). India's urban revolution: Geographies of displacement beyond gentrification. *Environment and Planning A, 46*, 1554–1571.

Ghertner, D. A. (2015). *Rule by Aesthetics: World-class City Making in Delhi.* Oxford: Oxford University Press.

Ghertner, D. A., & Lake, R. W. (2021). *Land Fictions: The Commodification of Land in City and Country*: Ithaca: Cornell University Press.

Gibson, C., Legacy, C., & Rogers, D. (2023). Deal-making, elite networks and public–private hybridisation: More-than-neoliberal urban governance. *Urban Studies, 60*, 183–199.

Gilbert, A. (2007). The return of the slum: Does language matter? *International Journal of Urban and Regional Research, 31*, 697–713.

Glass, R. (1964). *London: Aspects of change.* London: Centre for Urban Studies, University College London.

Glassman, J., & Choi, Y.-J. (2014). The Chaebol and the US military–industrial complex: Cold War geopolitical economy and South Korean industrialization. *Environment and Planning A, 46*, 1160–1180.

Goldman, M. (2020). Dispossession by financialization: The end(s) of rurality in the making of a speculative land market. *The Journal of Peasant Studies, 47*, 1251–1277.

Goldman, M. (2021). Speculative urbanism and the urban-financial conjuncture: Interrogating the afterlives of the financial crisis. *Environment and Planning A: Economy and Space*, https://doi.org/10.1177/0308518X211016003

Gong, H., Hassink, R., & Wang, C. C. (2022). Strategic coupling and institutional innovation in times of upheavals: The industrial chain chief model in Zhejiang, China. *Cambridge Journal of Regions, Economy and Society, 15*, 279–303.

Gonzalez, S. (2020). Contested marketplaces: Retail spaces at the global urban margins. *Progress in Human Geography, 44*, 877–897.

Goron, C. (2018). Ecological civilisation and the political limits of a Chinese concept of sustainability. *China Perspectives, 2018*, 39–52.

Gotham, K. F. (2009). Creating liquidity out of spatial fixity: The secondary circuit of capital and the subprime mortgage crisis. *International Journal of Urban and Regional Research, 33*, 355–371.

Grimes, S., & Miozzo, M. (2015). Big Pharma's internationalization of R&D to China. *European Planning Studies, 23*, 1873–1894.

Guironnet, A., Attuyer, K., & Halbert, L. (2016). Building cities on financial assets: The financialisation of property markets and its implications for city governments in the Paris city-region. *Urban Studies, 53*, 1442–1464.

Haila, A. (1999). City building in the East and West. *Cities, 16*, 259–267.

Haila, A. (2000). Real Estate in Global Cities: Singapore and Hong Kong as Property States. *Urban Studies, 37*, 2241–2256.

Haila, A. (2016). *Urban Land Rent: Singapore As a Property State*. Oxford: Wiley-Blackwell.

Halbert, L., & Rouanet, H. (2014). Filtering risk away: Global finance capital, transcalar territorial networks and the (un)making of city-regions: An analysis of business property development in Bangalore, India. *Regional Studies, 48*, 471–484.

Hall, P. A., & Soskice, D. (2001). *Varieties of Capitalism: The Institutional Foundations of Comparative Advantage*. Oxford: Oxford University Press.
Hamel, P., & Keil, R. (2015). *Suburban Governance: A Global View*. Toronto: University of Toronto Press.
Hamnett, C. (2020). Is Chinese urbanisation unique? *Urban Studies*, *57*, 690–700.
Hamnett, C. (2021). *Advanced Introduction to Gentrification*. Cheltenham, UK: Edward Elgar Publishing.
Hamnett, C., & Shoval, N. (2003). Museums as flagships of urban development. In L. M. Hoffman, S. Fainstein, & D. R. Judd (Eds.), *Cities and Visitors: Regulating People, Markets, and City Space* (pp. 219–236). Oxford: John Wiley & Sons.
Harrison, P., Rubin, M., Appelbaum, A., & Dittgen, R. (2019). Corridors of freedom: Analyzing Johannesburg's ambitious inclusionary transit-oriented development. *Journal of Planning Education and Research*, *39*, 456–468.
Harvey, D. (1978). The urban process under capitalism. *International Journal of Urban and Regional Research*, *2*, 101–131.
Harvey, D. (1982). *The Limits to Capital*. Oxford: Blackwell.
Harvey, D. (1989). From managerialism to entrepreneurialism: The transformation in urban governance in late capitalism. *Geografiska Annaler*, *71B*, 3–17.
Harvey, D. (2003). *The New Imperialism*. Oxford: Oxford University Press.
Harvey, D. (2005). *A Brief History of Neoliberalism*. Oxford: Oxford University Press.
Harvey, D. (2007). Neoliberalism as creative destruction. The Annals of American Academy of Political and Social Sciences, 610, 22–44.
Harvey, D. (2016). *The Ways of the World*. London: Profile Books.
Harvey, D. (2023). *A Companion to Marx's Grundrisse*. London: Verso.
He, S., & Qian, J. (2023). Police and politics in aesthetics-based urban governance: redevelopment and grassroots struggles in Enninglu, Guangzhou, China. *Antipode*, *55*, 853–876.
He, S., Wang, D., Webster, C., & Chau, K. W. (2019). Property rights with price tags? Pricing uncertainties in the production, transaction and consumption of China's small property right housing. *Land Use Policy*, *81*, 424–433.
He, S., & Wu, F. (2005). Property-led redevelopment in post-reform China: A case study of Xintiandi redevelopment project in Shanghai. *Journal of Urban Affairs*, *27*, 1–23.
He, S., & Wu, F. (2009). China's emerging neoliberal urbanism: Perspectives from urban redevelopment. *Antipode*, *41*, 282–304.
He, S., Zhang, M., & Wei, Z. (2020). The state project of crisis management: China's Shantytown Redevelopment Schemes under state-led financialization. *Environment and Planning A: Economy and Space*, *52*, 632–653.
Heilmann, S. (2008). Policy experimentation in China's economic rise. *Studies in Comparative International Development*, *43*, 1–26.
Heilmann, S., Shih, L., & Hofem, A. (2013). National planning and local technology zones: Experimental governance in China's Torch Programme. *The China Quarterly*, *216*, 896–919.
Hilbrandt, H. (2017). Insurgent participation: Consensus and contestation in planning the redevelopment of Berlin-Tempelhof airport. *Urban Geography*, *38*, 537–556.
Hirt, S. A. (2005). Planning the post-communist city: Experience from Sofia. *International Planning Studies*, *10*, 219–240.
Hirt, S. A. (2012). *Iron Curtains: Gates, Suburbs and Privatization of Space in the Post-socialist CityW*. Oxford: Wiley-Blackwell.
Hofman, A., & Aalbers, M. B. (2019). A finance-and real estate-driven regime in the United Kingdom. *Geoforum*, *100*, 89–100.
Hsing, Y.-T. (2010). *The Great Urban Transformation: Politics of Land and Property in China*. Oxford: Oxford University Press.
Hsu, J. Y., & Saxenian, A. (2000). The limits of guanxi capitalism: Transnational collaboration between Taiwan and the USA. *Environment and Planning A*, *32*, 1991–2005.
Hu, X., & Hassink, R. (2017). Place leadership with Chinese characteristics? A case study of the Zaozhuang coal-mining region in transition. *Regional Studies*, *51*, 224–234.
Huang, P., Westman, L., & Castan Broto, V. (2021). A culture-led approach to understanding energy transitions in China: The correlative epistemology. *Transactions of the Institute of British Geographers*, *46*, 900–916.
Huang, Y. (2008). *Capitalism with Chinese Characteristics: Entrepreneurship and the State*. Cambridge: Cambridge University Press.

Huang, Y. (2020). *Chinese Cities in the 21st Century*. New York: Springer.

Hubbard, P. (2018). Retail gentrification. In L. Lees & M. Phillips (Eds.), *Handbook of Gentrification Studies* (pp. 294–309). Cheltenham, UK: Edward Elgar Publishing.

Hult, A. (2015). The circulation of Swedish urban sustainability practices: To China and back. *Environment and Planning A*, *47*, 537–553.

Imrie, R., & Thomas, H. (1999). *British Urban Policy: An Evaluation of the Urban Development Corporations*. London: Sage.

Ioannou, S., & Wójcik, D. (2019). On financialization and its future. *Environment and Planning A: Economy and Space*, *51*, 263–271.

Jacobs, J. (1970) *The Economy of Cities*. New York: Knopf Doubleday Publishing Group.

Jayne, M. (2018). *Chinese Urbanism: Critical Perspectives*. Abingdon, Oxon: Routledge.

Jayne, M., & Ward, K. (2016). *Urban Theory: New Critical Perspectives*. London: Routledge.

Jessop, B. (2002a). *The Future of the Capitalist State*. Cambridge: Polity Press.

Jessop, B. (2002b). Liberalism, neoliberalism, and urban governance: A state-theoretical perspective. *Antipode*, *34*, 452–472.

Jessop, B. (2006). Spatial fixes, temporal fixes and spatio-temporal fixes. In N. Castree & D. Gregory (Eds.), *David Harvey: A Critical Reader* (pp. 142–166). Oxford: Blackwell Publishing Ltd.

Jessop, B. (2016). *The State: Past, Present, Future*. Cambridge: Polity.

Jessop, B., & Sum, N. L. (2000). An entrepreneurial city in action: Hong Kong's emerging strategies in and for (inter)urban competition. *Urban Studies*, *37*, 2287–2313.

Jiang, Y., & Waley, P. (2022). Financialization of urban development in China: Fantasy, fact or somewhere in between? *Regional Studies*, *56*, 1271–1281.

Jonas, A. E. (2020). China's urban development in context: Variegated geographies of city-regionalism and managing the territorial politics of urban development. *Urban Studies*, *57*, 701–708.

Kaika, M., & Ruggiero, L. (2016). Land financialization as a 'lived'process: The transformation of Milan's Bicocca by Pirelli. *European Urban and Regional Studies*, *23*, 3–22.

Kaika, M., Varvarousis, A., Demaria, F., & March, H. (2023). Urbanizing degrowth: Five steps towards a Radical Spatial Degrowth Agenda for planning in the face of climate emergency. *Urban Studies*, *60*, 1191–1211.

Karwowski, E. (2019). Towards (de-)financialisation: The role of the state. *Cambridge Journal of Economics*, *43*, 1001–1027.

Kay, K., & Tapp, R. (2022). Un/making assets: The institutional limits to financialization. *Annals of the American Association of Geographers*, *112*, 1243–1259.

Keil, R. (2020). An urban political ecology for a world of cities. *Urban Studies*, *57*, 2357–2370.

Keil, R., & Wu, F. (2022). *After Suburbia: Urbanization on the Planet's Periphery*. Toronto: University of Toronto Press.

Keith, M., Lash, S., Arnoldi, J., & Rooker, T. (2014). *China Constructing Capitalism: Economic Life and Urban Change*. London: Routledge.

Kennedy, L. (2015). The politics and changing paradigm of megaproject development in metropolitan cities. *Habitat International*, *45*, 163–168.

Kennedy, L., & Sood, A. (2016). Greenfield development as Tabula Rasa: Rescaling, speculation and governance on India's urban frontier. *Economic & Political Weekly*, *51*, 41.

Kinossian, N., & Morgan, K. (2023). Authoritarian state capitalism: Spatial planning and the megaproject in Russia. *Environment and Planning A: Economy and Space*, *55*, 655–672.

Klink, J., Empinotti, V. L., & Aversa, M. (2020). On contested water governance and the making of urban financialisation: Exploring the case of metropolitan São Paulo, Brazil. *Urban Studies*, *57*, 1676–1695.

Kornai, J. (1980). *Economics of Shortage*. Amsterdam: North-Holland.

Kostka, G., & Nahm, J. (2017). Central–local relations: Recentralization and environmental governance in China. *The China Quarterly*, *231*, 567–582.

Kostka, G., & Zhang, C. (2018). Tightening the grip: Environmental governance under Xi Jinping. *Environmental Politics*, *27*, 769–781.

Krippner, G. R. (2005). The financialization of the American economy. *Socio-economic Review*, *3*, 173–208.

Lake, R. W. (2016). The subordination of urban policy in the time of financialization. In J. DeFilippis (Ed.), *Urban Policy in the Time of Obama* (pp. 45–64). Minneapolis, MN: University of Minnesota Press.

Lauermann, J. (2018). Municipal statecraft: Revisiting the geographies of the entrepreneurial city. *Progress in Human Geography*, *42*, 205–224.
Lawhon, M., & Murphy, J. T. (2012). Socio-technical regimes and sustainability transitions: Insights from political ecology. *Progress in Human Geography*, *36*, 354–378.
Le Galès, P. (2016). Neoliberalism and urban change: Stretching a good idea too far? *Territory, Politics, Governance*, *4*, 154–172.
Lee, C. K., & Zhang, Y. (2013). The power of instability: Unraveling the microfoundations of bargained authoritarianism in China. *American Journal of Sociology*, *118*, 1475–1508.
Lees, L., Shin, H. B., & López-Morales, E. (2016). *Planetary Gentrification*. Cambridge: Polity.
Legacy, C. (2016). Transforming transport planning in the post-political era. *Urban Studies*, *53*, 3108–3124.
Lehrer, U., & Laidley, J. (2008). Old mega-projects newly packaged? Waterfront redevelopment in Toronto. *International Journal of Urban and Regional Research*, *32*, 786–803.
Lehrer, U. A. (1994). Images of the periphery: The architecture of Flex Space in Switzerland. *Environment and Planning D: Society and Space*, *12*, 187–205.
Li, B., Tong, D., Wu, Y., & Li, G. (2021). Government-backed 'laundering of the grey'in upgrading urban village properties: Ningmeng apartment project in Shuiwei village, Shenzhen, China. *Progress in Planning*, *146*, 100436.
Li, H., & Zhou, L.-A. (2005). Political turnover and economic performance: The incentive role of personnel control in China. *Journal of Public Economics*, *89*, 1743–1762.
Li, J., & Chiu, R. L. H. (2020). State rescaling and large-scale urban development projects in China: The case of Lingang New Town, Shanghai. *Urban Studies*, *57*, 2564–2581.
Li, S.-M. (2010). Mortgage loan as a means of home finance in urban China: A comparative study of Guangzhou and Shanghai. *Housing Studies*, *25*, 857–876.
Li, Y., Chen, W., & Wu, F. (2023a). Building Chinese city-regions under state entrepreneurialism. *Territory, Politics, Governance*, 2023, 1–18.
Li, Y., & Shapiro, J. (2020). *China Goes Green: Coercive Environmentalism for a Troubled Planet*. Cambridge: Polity Press.
Li, Y., & Wu, F. L. (2012). The transformation of regional governance in China: The rescaling of statehood. *Progress in Planning*, *78*, 55–99.
Li, Z. (2022). The new growth machine and neighborhood redevelopment in the Chinese cities of the Xijinping era: The case of Jinsong, Beijing. *Transactions in Planning and Urban Research*, *1*, 235–250.
Li, Z., Wu, F., & Zhang, F. (2022). A multi-scalar view of urban financialization: Urban development and local government bonds in China. *Regional Studies*, *56*, 1282–1294.
Li, Z., Wu, F., & Zhang, F. (2023b). State de-financialisation through incorporating local government bonds in the budgetary process in China. *Journal of Economic Geography*, *23*, 1169–1190.
Li, Z., Wu, F., & Zhang, F. (2024). The political economy of China's local debt. *The China Quarterly*, *257*, 100–118.
Liao, F. (2015). Quenching thirst with poison? Local government financing vehicles: Past, present, and future. In B. L. Liebman & C. J. Milhaupt (Eds.), *Regulating the Visible Hand?* (pp. 69–84). Oxford: Oxford University Press.
Lieberthal, K. G., & Lampton, D. M. (2012). *Bureaucracy, Politics, and Decision Making in Post-Mao China*. Berkeley, CA: University of California Press.
Lim, K. F. (2014). 'Socialism with Chinese characteristics': Uneven development, variegated neoliberalization and the dialectical differentiation of state spatiality. *Progress in Human Geography*, *38*, 221–247.
Lim, K. F. (2019). *On Shifting Foundations: State Rescaling, Policy Experimentation and Economic Restructuring in Post-1949 China*. Oxford: Wiley.
Lin, G. C. S. (2014). China's landed urbanization: Neoliberalizing politics, land commodification, and municipal finance in the growth of metropolises. *Environment and Planning A*, *46*, 1814–1835.
Lin, G. C. S. (2015). The redevelopment of China's construction land: Practising land property rights in cities through renewals. *The China Quarterly*, *224*, 865–887.
Lin, G. C. S., & Ho, S. (2005). The state, land system, and land development processes in contemporary China. *Annals of Association of American Geographers*, *95*, 411–436.
Lin, G. C. S., & Kao, S. Y. (2020). Contesting eco-urbanism from below: The construction of "zero-waste neighborhoods" in Chinese cities. *International Journal of Urban and Regional Research*, *44*, 72–89.

Lin, G. C. S., & Yi, F. (2011). Urbanization of capital or capitalization on urban land? Land development and local public finance in urbanizing China. *Urban Geography*, *32*, 50–79.
Lincoln, T. (2021). *An Urban History of China*. Cambridge: Cambridge University Press.
Liu, A. Y., Oi, J. C., & Zhang, Y. (2022). China's local government debt: The grand bargain. *The China Journal*, *87*, 40–71.
Liu, M., & Lo, K. (2021). Governing eco-cities in China: Urban climate experimentation, international cooperation, and multilevel governance. *Geoforum*, *121*, 12–22.
Liu, M., & Lo, K. (2022). Eco-cities as urban laboratories of Chinese ecological modernization: Eco-experimentation and the modernization of urban governance, *44*, *Urban Geography*, 1–25.
Liu, R., Wong, T.-C., & Liu, S. (2012). Peasants' counterplots against the state monopoly of the rural urbanization process: Urban villages and 'small property housing'in Beijing, China. *Environment and Planning A*, *44*, 1219–1240.
Liu, X., & White, S. (2001). Comparing innovation systems: A framework and application to China's transitional context. *Research Policy*, *30*, 1091–1114.
Liu, Y., Li, Z., Liu, Y., & Chen, H. (2015). Growth of rural migrant enclaves in Guangzhou, China: Agency, everyday practice and social mobility. *Urban Studies*, *52*, 3086–3105.
Liu, Y., & Yau, Y. (2020). Urban entrepreneurialism vs market society: The geography of China's neoliberal urbanism. *International Journal of Urban and Regional Research*, *44*, 266–288.
Liu, Z. (2019). Land-based finance and property tax in China. *Area Development and Policy*, *4*, 367–381.
Liu, Z., Huang, Z., Yin, Z., & Zhang, L. (2023). Urban Regeneration under National Land Use Control: Guangdong's "Three-Old" Redevelopment Programme, *258*, *The China Quarterly*, 1–16.
Lo, K. (2023). The politics of climate experimentalism in China. In F. Zhang & F. Wu (Eds.), *Handbook on China's Urban Environmental Governance* (pp. 144–155). Cheltenham, UK: Edward Elgar.
Lo, K., Li, H., & Chen, K. (2020). Climate experimentation and the limits of top-down control: Local variation of climate pilots in China. *Journal of Environmental Planning and Management*, *63*, 109–126.
Logan, J. R. (2008). *Urban China in Transition*. Oxford: Blackwell.
Logan, J. R., & Molotch, H. L. (1987). *Urban Fortunes: The Political Economy of Place*. Berkeley: University of California Press.
Lora-Wainwright, A. (2017). *Resigned Activism: Living with Pollution in Rural China*. Cambridge, MA.: MIT Press.
Lord, A., & Tewdwr-Jones, M. (2014). Is Planning "Under Attack"? Chronicling the Deregulation of Urban and Environmental Planning in England. *European Planning Studies*, *22*, 345–361.
Lord, A., & Tewdwr-Jones, M. (2018). Getting the planners off our backs: Questioning the post-political nature of English planning policy. *Planning Practice & Research*, *33*, 229–243.
Lovering, J. (1999). Theory Led by Policy: The Inadequacies of the 'New Regionalism' (Illustration from the Case of Wales). *International Journal of Urban and Regional Research*, *?*, 379–395.
Luo, Y., & Shen, J. (2022). Urban entrepreneurialism, metagovernance and 'space of innovation': Evidence from buildings for innovative industries in Shenzhen, China. *Cities*, *131*, 104067.
Lyu, L., Wu, W., Hu, H., & Huang, R. (2019). An evolving regional innovation network: Collaboration among industry, university, and research institution in China's first technology hub. *The Journal of Technology Transfer*, *44*, 659–680.
Ma, L. J. C. (2005). Urban administrative restructuring, changing scale relations and local economic development in China. *Political Geography*, *24*, 477–497.
Ma, L. J. C., & Wu, F. (2005). *Restructuring the Chinese City: Changing Society, Economy and Space*. London: Routledge.
MacKinnon, D. (2012). Beyond strategic coupling: Reassessing the firm-region nexus in global production networks. *Journal of Economic Geography*, *12*, 227–245.
MacLeod, G. (2001). Beyond soft institutionalism: Accumulation, regulation, and their geographical fixes. *Environment and Planning A*, *33*, 1145–1167.
MacLeod, G. (2011). Urban politics reconsidered: Growth machine to post-democratic city? *Urban Studies*, *48*, 2629–2660.
Mai, X., Xu, Y., & Liu, Y. (2023). Cultivating an alternative subjectivity beyond neoliberalism: Community gardens in urban China. *Annals of the American Association of Geographers*, *113*, 1–17.
Mazzucato, M. (2013). *The Entrepreneurial State: Debunking Public Vs. Private Sector Myths*. London: Anthem Press.

Mazzucato, M. (2018). Mission-oriented innovation policies: Challenges and opportunities. *Industrial and Corporate Change*, *27*, 803–815.
McCann, E., & Ward, K. (2011). *Mobile Urbanism: Cities and Policy Making in the Global Age*. Minneapolis, MN: Minnesota University Press.
McFarlane, C. (2011). Assemblage and critical urbanism. *City*, *15*, 204–224.
McGee, T. G., & Robinson, I. M. (1995). *The mega-urban regions of Southeast Asia*. Vancouver: UBC Press.
McGuirk, P. M. (2000). Power and policy networks in urban governance: Local government and property-led regeneration in Dublin. *Urban Studies*, *37*, 651–672.
Mertha, A. (2009). "Fragmented authoritarianism 2.0": Political pluralization in the Chinese policy process. *The China Quarterly*, *200*, 995–1012.
Metzger, J. (2017). Postpolitics and planning. In M. Gunder, A. Madanipour, & V. Watson (Eds.), *The Routledge Handbook of Planning Theory* (pp. 180–193). London: Routledge.
Miao, J. T., Kim, H. M., & Phelps, N. A. (2023). Competition and coordination in state intrapreneurialism: The case of South Korea's export of urban expertise. *Environment and Planning A: Economy and Space*, https://doi.org/10.1177/0308518X231182428
Miao, J. T., & Phelps, N. A. (2019). The intrapreneurial state: Singapore's emergence in the smart and sustainable urban solutions field. *Territory, Politics, Governance*, *7*, 316–335.
Miao, J. T., & Phelps, N. A. (2022). Urban sprawl as policy sprawl: Distinguishing Chinese capitalism's suburban spatial fix. *Annals of the American Association of Geographers*, *112*, 1179–1194.
Miles, S., & Paddison, R. (2005). Introduction: The rise and rise of culture-led urban regeneration. *Urban Studies*, *42*, 833–839.
Moisio, S., Koch, N., Jonas, A. E., Lizotte, C., & Luukkonen, J. (2020). *Handbook on the Changing Geographies of the State: New Spaces of Geopolitics*. Cheltenham, UK: Edward Elgar Publishing.
Moisio, S., & Rossi, U. (2020). The start-up state: Governing urbanised capitalism. *Environment and Planning A: Economy and Space*, *52*, 532–552.
Mouffe, C. (2005). *On the Political*. London: Routeldge.
Murray, M. J. (2015). Waterfall City (Johannesburg): Privatized urbanism in extremis. *Environment and Planning A*, *47*, 503–520.
Murray, M. J. (2017). *The Urbanism of Exception: The Dynamics of Global City Building in the Twenty-first Century*. Cambridge: Cambridge University Press.
Myers, G. (2020). *Rethinking Urbanism: Lessons from Postcolonialism and the Global South*. Bristol: Bristol University Press.
Naughton, B. (2020). Xi Jinping's economic policy and Chinese urbanization. In Y. Huang (Ed.), *Chinese Cities in the 21st Century*. Cham., UK: Palgrave Macmillan.
Naughton, B., & Tsai, K. S. (2015). *State capitalism, Institutional Adaptation, and the Chinese Miracle*. Cambridge: Cambridge University Press.
Oi, J. C. (1992). Fiscal reform and the economic foundations of local state corporatism. *World Politics*, *45*, 99–126.
Olds, K. (2001) *Globalization and urban change: capital, culture, and Pacific Rim mega-projects*. Oxford: Oxford University Press.
O'Mara, M. (2020). *The Code: Silicon Valley and the Remaking of America*. London: Penguin.
Ong, A. (2006). *Neoliberalism as Exception: Mutations in Citizenship and Sovereignty*. Durham: Duke University Press.
Ong, A. (2007). Neoliberalism as a mobile technology. *Transactions of the Institute of British Geographers*, *32*, 3–8.
Ong, A. (2011). Worlding cities, or the art of being global. In A. Roy & A. Ong (Eds.), *Worlding Cities: Asian Experiments and the Art of Being Global* (pp. 1–26). Oxford: Wiley-Blackwell.
Ong, L. H. (2014). State-led urbanization in China: Skyscrapers, land revenue and 'concentrated villages'. *The China Quarterly*, *217*, 162–179.
Orueta, F. D., & Fainstein, S. S. (2008). The new mega-projects: Genesis and impacts. *International Journal of Urban and Regional Research*, *32*, 759–767.
Pan, F., Zhang, F., & Wu, F. (2021). State-led financialization in China: The case of the government-guided investment fund. *The China Quarterly*, *247*, 749–772.
Pan, F., Zhang, F., Zhu, S., & Wójcik, D. (2017). Developing by borrowing? Inter-jurisdictional competition, land finance and local debt accumulation in China. *Urban Studies*, *54*, 897–916.
Pan, J. (2016). *China's Environmental Governing and Ecological Civilization*. New York: Springer.

Park, B.-G., Hill, R. C., & Saito, A. (2012). *Locating Neoliberalism in East Asia: Neoliberalizing Spaces in Developmental States*. Oxford: Wiley-Blackwell.

Parnell, S., & Pieterse, E. (2016). Translational global praxis: Rethinking methods and modes of African urban research. *International Journal of Urban and Regional Research*, *40*, 236–246.

Parnell, S., & Robinson, J. (2012). (Re) theorizing cities from the Global South: Looking beyond neoliberalism. *Urban Geography*, *33*, 593–617.

Pearson, M., Rithmire, M., & Tsai, K. (2023). *The State and Capitalism in China*. Cambridge: Cambridge University Press.

Peck, J. (2011). Neoliberal suburbanism: Frontier space. *Urban Geography*, *32*, 884–919.

Peck, J. (2012). Austerity urbanism: American cities under extreme economy. *City*, *16*, 626–655.

Peck, J. (2015). Cities beyond compare? *Regional Studies*, *49*, 160–182.

Peck, J. (2017a). Transatlantic city, part 1: Conjunctural urbanism. *Urban Studies*, *54*, 4–30.

Peck, J. (2017b). Transatlantic city, part 2: Late entrepreneurialism. *Urban Studies*, *54*, 327–363.

Peck, J. (2021). On capitalism's cusp. *Area Development and Policy*, *6*, 1–30.

Peck, J. (2023). Practicing conjunctural methodologies: Engaging Chinese capitalism. *Dialogues in Human Geography*. https://doi.org/10.1177/20438206231154346

Peck, J., & Tickell, A. (2002). Neoliberalizing space. *Antipode*, *34*, 380–404.

Peck, J., & Whiteside, H. (2016). Financializing Detroit. *Economic Geography*, *92*, 235–268.

Peck, J., & Zhang, J. (2013). A variety of capitalism … with Chinese characteristics? *Journal of Economic Geography*, *13*, 357–396.

Penny, J. (2022). 'Revenue generating machines'? London's local housing companies and the emergence of local state rentierism. *Antipode*, *54*, 545–566.

Petry, J. (2020). Financialization with Chinese characteristics? Exchanges, control and capital markets in authoritarian capitalism. *Economy and Society*, *49*, 213–238.

Phelps, N., & Wu, F. (2011). *International Perspectives on Suburbanization: A Post-suburban World?* Basingstoke: Palgrave Macmillan.

Phelps, N. A., & Miao, J. T. (2020). Varieties of urban entrepreneurialism. *Dialogues in Human Geography*, *10*, 304–321.

Pierre, J. (2011). *The Politics of Urban Governance*. Basingstoke, Hampshire: Palgrave Macmillan.

Pierre, J., & Peters, B. G. (2020). *Governance, Politics and the State* (2nd ed.). London: Bloomsbury Publishing.

Pike, A. (2023). *Financialization and Local Statecraft*. Oxford: Oxford University Press.

Pike, A., O'Brien, P., Strickland, T., & Tomaney, J. (2019). *Financialising City Statecraft and Infrastructure*. Cheltenham, UK: Edward Elgar Publishing.

Pinson, G., & Journel, C. M. (2017). *Debating the Neoliberal City*. London: Taylor & Francis.

Po, L. (2011). Property rights reforms and changing grassroots governance in China's urban-rural peripheries: The case of Changping District in Beijing. *Urban Studies*, *48*, 509–528.

Polanyi K (1944) *The Great Transformation: The Political Economic Origin of Our Time*. New York: Farrar & Rinehart

Porter, M. E. (2000). Location, competition, and economic development: Local clusters in a global economy. *Economic Development Quarterly*, *14*, 15–34.

Pow, C.-P. (2012). China exceptionalism? Unbounding narratives on urban China. In T. Edensor & M. Jayne (Eds.), *Urban Theory Beyond the West: A World of Cities* (pp. 62–79). Abingdon: Routledge.

Qian, J., & He, S. (2012). Rethinking social power and the right to the city amidst China's emerging urbanism. *Environment and Planning A*, *44*, 2801–2816.

Qian, J., He, S., & Liu, L. (2013). Aestheticisation, rent-seeking, and rural gentrification amidst China's rapid urbanisation: The case of Xiaozhou village, Guangzhou. *Journal of Rural Studies*, *32*, 331–345.

Qian, Z. (2022). *Property Rights and Urban Transformation in China*. Cheltenham, UK: Edward Elgar Publishing.

Qin, H. (2008). The formation and future of China miracle: A personal view of China's thirty years' reform. *Nanfang Zoumo (Southern Weekends), 21 Feburary 2008, p5.*

Raco, M. (2014). Delivering flagship projects in an era of regulatory capitalism: State-led privatization and the London Olympics 2012. *International Journal of Urban and Regional Research*, *38*, 176–197.

Rancière, J. (1999). *Disagreement: Politics and Philosophy*. Minneapolis, MN: University of Minnesota Press.

Rapoport, E., & Hult, A. (2017). The travelling business of sustainable urbanism: International consultants as norm-setters. *Environment and Planning A: Economy and Space*, *49*, 1779–1796.

Register, R. (1987). *Ecocity Berkeley: Building cities for a healthy future. Berkeley, CA North Atlantic Books*. Berkeley, CA: North Atlantic Books is Publisher.
Ren, J. (2015). Gentrification in China? In L. Lees, H. B. Shin, & E. Lopez-Morales (Eds.), *Global Gentrifications: Uneven Development and Displacement* (pp. 335–354). Bristol: Policy Press.
Ren, J. (2023). Exceptionalism and theorizing spatial inequality: Segregation research on cities in China. *Journal of Urban Affairs*, *45*, 1225–1237.
Ren, X. (2008). Forward to the past: Historical preservation in globalizing Shanghai. *City and Community*, *7*, 23–43.
Ren, X. (2013). *Urban China*. Cambridge: Polity Press.
Rhodes, R. A. W. (1996). The new governance: Governing without government. *Political Studies*, *44*, 652–667.
Risager, B. S. (2021). Financialized gentrification and class composition in the post-industrial city: A rent strike against a real estate investment trust in Hamilton, Ontario. *International Journal of Urban and Regional Research*, *45*, 282–302.
Roberts, P., Sykes, H., & Granger, R. (2016). *Urban Regeneration* (2nd ed.). London: SAGE Publications.
Robinson, J. (2022). *Comparative Urbanism: Tactics for Global Urban Studies*. Oxford: John Wiley & Sons.
Robinson, J., & Attuyer, K. (2021). Extracting value, London style: Revisiting the role of the state in urban development. *International Journal of Urban and Regional Research*, *45*, 303–331.
Robinson, J., Harrison, P., Shen, J., & Wu, F. (2021). Financing urban development, three business models: Johannesburg, Shanghai and London. *Progress in Planning*, *154*, 100513.
Robinson, J., Wu, F., Harrison, P., Wang, Z., Todes, A., Dittgen, R., & Attuyer, K. (2022). Beyond variegation: The territorialisation of states, communities and developers in large-scale developments in Johannesburg, Shanghai and London. *Urban Studies*, *59*, 1715–1740.
Rodenbiker, J. (2020). Urban ecological enclosures: Conservation planning, peri-urban displacement, and local state formations in China. *International Journal of Urban and Regional Research*, *44*, 691–710.
Rodenbiker, J. (2022). Social justice in China's cities: Urban-rural restructuring and justice-oriented planning. *Transactions in Planning and Urban Research*, *1*, 184–198.
Rossi, U. (2017). *Cities in Global Capitalism*. Cambridge: John Wiley & Sons.
Roth, L., Russell, B., & Thompson, M. (2023). Politicising proximity: Radical municipalism as a strategy in crisis. *Urban Studies*, 60, 2009–2035.
Roy, A. (2005). Urban informality: Towards an epistemology of planning. *Journal of the American Planning Association 71*, 147–158.
Roy, A. (2011). The blockade of the world-class city: Dialectical images of Indian urbanism. In A. Roy & A. Ong (Eds.), *Worlding Cities: Asian Experiments and the Art of Being Global* (pp. 259–278). Oxford: Wiley-Blackwell.
Roy, A. (2015). Governing the postcolonial suburbs. In P. Hamel & R. Keil (Eds.), *Suburban Governance: A Global View* (pp. 337–347). Toronto: Toronto University Press.
Roy, A., & Ong, A. (2011). *Worlding Cities: Asian Experiments and the Art of Being Global*. Oxford: Wiley-Blackwell.
Rutland, T. (2010). The financialization of urban redevelopment. *Geography Compass*, *4*, 1167–1178.
Sa, H., & Haila, A. (2023). Urban villagers as real estate developers: Embracing property mind through 'planting'housing in North-east China. *Housing Studies*, *38*, 423–443.
Sager, T. (2011). Neo-liberal urban planning policies: A literature survey 1990–2010. *Progress in Planning*, *76*, 147–199.
Saich, T. (2015). *Governance and Politics of China* (4th ed.). London: Macmillan.
Scott, A. J., & Storper, M. (2015). The nature of cities: The scope and limits of urban theory. *International Journal of Urban and Regional Research*, *39*, 1–15.
Shao, Q. (2013). *Shanghai gone: Domicide and defiance in a Chinese megacity*. Lanham, MD: Rowman & Littlefield Publishers.
Shapiro, J. (2016). *China's Environmental Challenges*. Cambridge: Polity.
Shatkin, G. (2016). The real estate turn in policy and planning: Land monetization and the political economy of peri-urbanization in Asia. *Cities*, *53*, 141–149.
Shatkin, G. (2017). *Cities for Profit: The Real Estate Turn in Asia's Urban Politics*. Ithaca: NY: Cornell University Press.
Shearmur, R. (2012). Are cities the font of innovation? A critical review of the literature on cities and innovation. *Cities*, *29*, S9–S18.

Shen, J. (2020). Universities as financing vehicles of (sub) urbanisation: The development of university towns in Shanghai. *Land Use Policy*, *112*, 104679.

Shen, J., Luo, X., & Sun, Z. (2022). Housing (de-) financialisation under state entrepreneurialism in China: The revival of affordable housing and Shanghai's shared-ownership housing scheme. *Transactions in Planning and Urban Research*, *1*, 269–288.

Shen, J., Luo, X., & Wu, F. (2020). Assembling mega-urban projects through state-guided governance innovation: The development of Lingang in Shanghai. *Regional Studies*, *54*, 1644–1654.

Shen, J., & Wu, F. (2020). Paving the way to growth: Transit-oriented development as a financing instrument for Shanghai's post-suburbanization. *Urban Geography*, *41*, 1010–1032.

Shi, C., & Tang, B.-S. (2020). Institutional change and diversity in the transfer of land development rights in China: The case of Chengdu. *Urban Studies*, *57*, 473–489.

Shi, Y. (2023). Is China financialised? The significance of two historic transformations of Chinese finance. *New Political Economy*, *29*, 1–16.

Shin, H. B. (2016). Economic transition and speculative urbanisation in China: Gentrification versus dispossession. *Urban Studies*, *53*, 471–489.

Shue, V., & Thornton, P. M. (2017). *To govern China: Evolving practices of power*. Cambridge: Cambridge University Press.

Sit, V. F. S., & Yang, C. (1997). Foreign-investment-induced exo-urbanisation in the Pearl River Delta, China. *Urban Studies*, *34*, 647–677.

Smart, A. (2018). Ethnographic perspectives on the mediation of informality between people and plans in urbanising China. *Urban Studies*, *55*, 1477–1483.

Smart, A., & Lee, J. (2003). Financialization and the role of real estate in Hong Kong's regime of accumulation. *Economic Geography*, *79*, 153–171.

Smith, N. (2002). New globalism, new urbanism: Gentrification as global urban strategy. *Antipode*, *34*, 427–450.

Solinger, D. J. (2002). Labour market reform and the plight of the laid-off proletariat. *The China Quarterly*, *170*, 304–326.

Solinger, D. J. (2018). *Polarized cities: Portraits of rich and poor in urban* China: Rowman & Littlefield.

Sood, A., & Kennedy, L. (2020). Neoliberal exception to liberal democracy? Entrepreneurial territorial governance in India. *Territory, Politics, Governance*, *8*, 23–42.

Storper, M., & Scott, A. J. (2016). Current debates in urban theory: A critical assessment. *Urban Studies*, *53*, 1114–1136.

Storper, M., & Venables, A. J. (2004). Buzz: Face-to-face contact and the urban economy. *Journal of Economic Geography*, *4*, 351–370.

Su, F., & Tao, R. (2017). The China model withering? Institutional roots of China's local developmentalism. *Urban Studies*, *54*, 230–250.

Su, F., & Tao, R. (2020). Urbanism in Chinese local development: An institutional approach. *Chinese Cities in the 21st Century* (pp. 93–116): Palgrave Macmillan, Cham.

Su, X. (2015). Urban entrepreneurialism and the commodification of heritage in China. *Urban Studies*, *52*, 2874–2889.

Su, X. (2022). Unpacking administrative rank: Intercity competition and the remaking of local state space in China. *Political Geography*, *92*, 102518.

Sun, Y., Ling, J., & He, S. (2023). Actually existing state entrepreneurialism: From conceptualization to materialization. *Progress in Human Geography*. https://doi.org/10.1177/03091325231219700

Swyngedouw, E. (2009). The antinomies of the postpolitical city: In search of a democratic politics of environmental production. *International Journal of Urban and Regional Research*, *33*, 601–620.

Swyngedouw, E. (2018). *Promises of the Political: Insurgent Cities in a Post-political Environment*. Cambridge, MA.: The MIT Press.

Swyngedouw, E., Moulaert, F., & Rodriguez, A. (2002). Neoliberal urbanization in Europe: Large-scale urban development projects and the new urban policy. *Antipode*, *34*, 542–577.

Tallon, A. (2021). *Urban Regeneration in the UK* (3rd ed.). London: Routledge.

Tao, R., Su, F., Liu, M., & Cao, G. (2010). Land leasing and local public finance in China's regional development: Evidence from prefecture-level cities. *Urban Studies*, *47*, 2217–2236.

Témenos, C., & McCann, E. (2012). The local politics of policy mobility: Learning, persuasion, and the production of a municipal sustainability fix. *Environment and Planning A*, *44*, 1389–1406.

Teo, S. S., Chung, C. K. L., & Wang, Z. (2023). Theorising with urban China: Methodological and tactical experiments for a more global urban studies. *Dialogues in Human Geography*. https://doi.org/10.1177/20438206231156656

Theurillat, T., & Graezer Bideau, F. (2022). China's extended urbanization driven by the "consumption city" in the context of financialized ecological civilization. *Transactions in Planning and Urban Research*, *1*, 17–31.

Theurillat, T., Vera-Büchel, N., & Crevoisier, O. (2016). Commentary: From capital landing to urban anchoring: The negotiated city. *Urban Studies*, *53*, 1509–1518.

Thompson, M. (2021). What's so new about New Municipalism? *Progress in Human Geography*, *45*, 317–342.

Thompson, M. (2023). Whatever happened to municipal radicalism? *Transactions of the Institute of British Geographers*, *48*, 603–618.

Thompson, M., Nowak, V., Southern, A., Davies, J., & Furmedge, P. (2020). Re-grounding the city with Polanyi: From urban entrepreneurialism to entrepreneurial municipalism. *Environment and Planning A: Economy and Space*, *52*, 1171–1194.

Tian, L., & Guo, Y. (2019). *Peri-Urban China: Land Use, Growth, and Integrated Urban–Rural Development*. London: Routledge.

Tong, D., Yang, S., & Lai, Y. (2023). Experimentation-based Policymaking for Urban Regeneration in Shenzhen, China. *The China Quarterly*, *258*, 1–18.

Tsui, K. Y. (2011). China's infrastructure investment boom and local debt crisis. *Eurasian Geography and Economics*, *52*, 686–711.

Turok, I. (1992). Property-led urban regeneration: Panacea or placebo? *Environment and Planning A*, *24*, 361–379.

Uitermark, J. (2014). Integration and Control: The Governing of Urban Marginality in Western Europe. *International Journal of Urban and Regional Research*, *38*, 1418–1436.

Uitermark, J., & Nicholls, W. (2014). From politicization to policing: The rise and decline of new social movements in Amsterdam and Paris. *Antipode*, *46*, 970–991.

Van Loon, J., Oosterlynck, S., & Aalbers, M. B. (2019). Governing urban development in the Low Countries: From managerialism to entrepreneurialism and financialization. *European Urban and Regional Studies*, *26*, 400–418.

Vicario, L., & Martínez Monje, P. M. (2003). Another 'Guggenheim effect'? The generation of a potentially gentrifiable neighbourhood in Bilbao. *Urban Studies*, *40*, 2383–2400.

Walder, A. G. (1995). Local governments as industrial firms: An organizational analysis of China's transitional economy. *American Journal of Sociology*, *101*, 263–301.

Walker, R., & Buck, D. (2007). The Chinese road: Cities in the transition to capitalism. *New Left Review*, *46*, 39–46.

Wang, J., & Li, S. (2017). State territorialization, neoliberal governmentality: The remaking of Dafen oil painting village, Shenzhen, China. *Urban Geography*, *38*, 708–728.

Wang, K., Qian, J., & He, S. (2021). Contested worldings of e-waste environmental justice: Nonhuman agency and e-waste scalvaging in Guiyu, China. *Annals of the American Association of Geographers*, *111*, 2165–2184.

Wang, K., Qian, J., & He, S. (2023a). From state entrepreneurialism to state-led ecological civilisation: Changing dispositifs of governing e-waste metabolism and 'cyborg' urbanisation in China's e-waste cities. In F. Zhang & F. Wu (Eds.), *Handbook on China's Urban Environmental Governance* (pp. 323–339). Cheltenham, UK: Edward Elgar.

Wang, L. (2022). China's new town movements since 1949: A state/space perspective. *Progress in Planning*, *155*, 100514.

Wang, M., Zhang, F., & Wu, F. (2022a). Governing urban redevelopment: A case study of Yongqingfang in Guangzhou, China. *Cities*, *120*, 103420.

Wang, M., Zhang, F., & Wu, F. (2022b). "Micro-regeneration": Toward small-scale, heritage-oriented, and participatory redevelopment in China. *Journal of Urban Affairs*, 1–18. https://doi.org/10.1080/07352166.2022.2139711

Wang, W., Wu, F., & Zhang, F. (2023b). Environmental city-regionalism in China: War against air pollution in the Beijing-Tianjin-Hebei Region. *Transactions in Planning and Urban Research*, *2*, 132–148.

Wang, Y. (2015). The rise of the 'shareholding state': Financialization of economic management in China. *Socio-economic Review*, *13*, 603–625.

Wang, Y. (2020). Financialization and state transformations. In P. Mader, D. Mertens, & N. V. D. Zwan (Eds.), *The Routledge International Handbook of Financialization* (pp. 118–197). London: Routledge.

Wang, Z., & Wu, F. (2019). In-situ marginalisation: Social impact of Chinese mega-projects. *Antipode, 51*, 1640–1663.
Ward, C., & Swyngedouw, E. (2018). Neoliberalisation from the ground up: Insurgent capital, regional struggle, and the assetisation of land. *Antipode, 50*, 1077–1097.
Ward, C., Van Loon, J., & Wijburg, G. (2019). Neoliberal Europeanisation, Variegated Financialisation: Common but divergent economic trajectories in the Netherlands, United Kingdom and Germany. *Tijdschrift voor Economische en Sociale Geografie, 110*, 123–137.
Ward, K., Jonas, A. E. G., Miller, B., & Wilson, D. (2018). *The Routledge Handbook on Spaces of Urban Politics: Spaces of Urban Politics*. London: Routledge.
Weber, R. (2010). Selling city futures: The financialization of urban redevelopment policy. *Economic Geography, 86*, 251–274.
Weinstein, L., & Ren, X. F. (2009). The Changing Right to the City: Urban Renewal and Housing Rights in Globalizing Shanghai and Mumbai. *City & Community, 8*, 407–432.
Westman, L. K., Broto, V. C., & Huang, P. (2019). Revisiting multi-level governance theory: Politics and innovation in the urban climate transition in Rizhao, China. *Political Geography, 70*, 14–23.
While, A., Jonas, A. E., & Gibbs, D. (2004). The environment and the entrepreneurial city: Searching for the urban 'sustainability; fix'in Manchester and Leeds. *International Journal of Urban and Regional Research, 28*, 549–569.
While, A., Jonas, A. E. G., & Gibbs, D. (2010). From sustainable development to carbon control: Eco-state restructuring and the politics of urban and regional development. *Transactions of the Institute of British Geographers, 35*, 76–93.
Whiteside, H. (2019). The state's estate: Devaluing and revaluing 'surplus' public land in Canada. *Environment and Planning A: Economy and Space, 51*, 505–526.
Whiteside, H. (2023). State and collective ownership: Thwarting and enabling financialization? *Urban Geography, 44*, 326–336.
Wijburg, G. (2021). The de-financialization of housing: Towards a research agenda. *Housing Studies, 36*, 1276–1293.
Wilson, J., & Swyngedouw, E. (2014). Seeds of dystopia: Post-politics and the return of the political. In: Wilson J and Swyngedouw E. (eds), *The Post-political and Its Discontents: Spaces of Depoliticisation, Spectres of Radical Politics*. Edinburgh: Edinburgh University Press, 1–22.
Wong, S. W., Chen, X., Tang, B.-S., & Liu, J. (2024). Neoliberal State Intervention and the Power of Community in Urban Regeneration: An Empirical Study of Three Village Redevelopment Projects in Guangzhou, China. *Journal of Planning Education and Research, 44*, 619–631.
Wu, F. (1997). Urban restructuring in China's emerging market economy: Towards a framework for analysis. *International Journal of Urban and Regional Research, 21*, 640–663.
Wu, F. (2000a). The global and local dimensions of place-making: Remaking Shanghai as a world city. *Urban Studies, 37*, 1359–1377.
Wu, F. (2000b). Place promotion in Shanghai, PRC. *Cities, 17*, 349–361.
Wu, F. (2002). China's changing urban governance in the transition towards a more market-oriented economy. *Urban Studies, 39*, 1071–1093.
Wu, F. (2003a. The (post-)socialist entrepreneurial city as a state project: Shanghai's reglobalisation in question. *Urban Studies, 40*, 1673–1698.
Wu, F. (2003b. Transitional cities. *Environment and Planning A, 35*, 1331–1338.
Wu, F. (2007). *China's Emerging Cities: The Making of New Urbanism*. London: Routledge.
Wu, F. (2010). How neoliberal is China's reform? The origins of change during transition. *Eurasian Geography and Economics, 51*, 619–631.
Wu, F. (2012). China's eco-cities. *Geoforum, 43*, 169–171.
Wu, F. (2015a). Commodification and housing market cycles in Chinese cities. *International Journal of Housing Policy, 15*, 6–26.
Wu, F. (2015b). *Planning for Growth: Urban and Regional Planning in China*. London: Routledge.
Wu, F. (2016a). China's emergent city-region governance: A new form of state spatial selectivity through state-orchestrated rescaling. *International Journal of Urban and Regional Research, 40*, 1134–1151.
Wu, F. (2016b). State dominance in urban redevelopment: Beyond gentrification in urban China. *Urban Affairs Review, 52*, 631–658.
Wu, F. (2017). State entrepreneurialism in urban China: A critique of the neoliberal city. In G. Pinson & C. M. Journel (Eds.), *Debating the Neoliberal City* (pp. 153–173). Abingdon, Oxon: Routledge.

Wu, F. (2018a). Housing privatization and the return of the state: Changing governance in China. *Urban Geography*, *39*, 1177–1194.
Wu, F. (2018b). Planning centrality, market instruments: Governing Chinese urban transformation under state entrepreneurialism. *Urban Studies*, *55*, 1383–1399.
Wu, F. (2020a). Adding new narratives to the urban imagination: An introduction to 'New directions of urban studies in China'. *Urban Studies*, *57*, 459–472.
Wu, F. (2020b). Scripting Indian and Chinese urban spatial transformation: Adding new narratives to gentrification and suburbanisation research. *Environment and Planning C*, *38*, 980–997.
Wu, F. (2022a). *Creating Chinese Urbanism: Urban Revolution and Governance Changes*. London: UCL Press.
Wu, F. (2022b). Intertwined modalities of suburban governance in China. In R. Keil & F. Wu (Eds.), *After Suburbia: Urbanization in the Twenty-first Century*. Toronto: University of Toronto Press.
Wu, F. (2022c). Land financialisation and the financing of urban development in China. *Land Use Policy*, *112*, 104412.
Wu, F. (2023a). The long shadow of the state: Financializing the Chinese city. *Urban Geography*, *44*, 37–58.
Wu, F. (2023b). State entrepreneurialism: Theorizing urban development politics from China. In P. Le Galès & J. Robinson (Eds.), *The Routledge Handbook of Comparative Global Urban Studies* (pp. 364–374). London: Routledge.
Wu, F. (2023c). Theorising urban development in China: 'state entrepreneurialism from the ground up. *Dialogues in Human Geography*. https://doi.org/10.1177/20438206231174638
Wu, F., Chen, J., Pan, F., Gallent, N., & Zhang, F. (2020). Assetisation: The Chinese path to housing financialisation. *Annals of the American Association of Geographers*, *110*, 1483–1499.
Wu, F., Deng, H., Feng, Y., Wang, W., Wang, Y. and Zhang, F., 2024. Statecraft at the frontier of capitalism: A grounded view from China. *Progress in Human Geography*. https://doi.org/10.1177/03091325241268953
Wu, F., & Keil, R. (2020). Changing the geographies of sub/urban theory: Asian perspectives. *Urban Geography*, *41*, 947–953. forthcoming.
Wu, F., & Logan, J. (2016). Do rural migrants 'float' in urban China? neighbouring and neighbourhood sentiment in Beijing. *Urban Studies*, *53*, 2973–2990.
Wu, F., Pan, F., & Chen, J. (2022a). Financialization under state entrepreneurialism in China. *Regional Studies*, *56*, 1237–1243.
Wu, F., & Phelps, N. A. (2011). (Post-)suburban development and state entrepreneurialism in Beijing's outer suburbs. *Environment and Planning A*, *43*, 410–430.
Wu, F., & Shen, J. (2015). Suburban development and governance in China. In P. Hamel & R. Keil (Eds.), *Suburban Governance: A Global View* (pp. 303–324). Toronto: University of Toronto Press.
Wu, F., Xu, J., & Yeh, A. G.-O. (2007). *Urban Development in Post-reform China: State,* Market and Space. London: Routledge.
Wu, F., & Zhang, F. (2022). Rethinking China's urban governance: The role of the state in neighbourhoods, cities and regions. *Progress in Human Geography*, *46*, 775–797.
Wu, F., Zhang, F., & Liu, Y. (2022b). Beyond growth machine politics: Understanding state politics and national political mandates in China's urban redevelopment. *Antipode*, *54*, 608–628.
Wu, F., Zhang, F., & Webster, C. (2013). Informality and the development and demolition of urban villages in the Chinese peri-urban area. *Urban Studies*, *50*, 1919–1934.
Wu, W. (2010). Managing and incentivizing research commercialization in Chinese Universities. *The Journal of Technology Transfer*, *35*, 203–224.
Wu, W., & Gao, Q. (2022). *China Urbanizing: Impacts and Transitions*. Philadelphia: University of Pennsylvania Press.
Wu, W., & Gaubatz, P. (2021). *The Chinese City* (2nd ed.). New York: Routledge.
Xie, L., Cheshmehzangi, A., Shao, M., Zhang, Y., & Chan, F. (2023). Eco-cities in China: National initiatives, local implementation and livelihood transitions. In F. Zhang & F. Wu (Eds.), *Handbook on China's Urban Environmental Governance* (pp. 169–185). Cheltenham, UK: Edward Elgar.
Xie, L., Cheshmehzangi, A., Tan-Mullins, M., Flynn, A., & Heath, T. (2020). Urban entrepreneurialism and sustainable development: A comparative analysis of Chinese eco-developments. *Journal of Urban Technology*, *27*, 3–26.
Xie, L., Flynn, A., Tan-Mullins, M., & Cheshmehzangi, A. (2019). The making and remaking of ecological space in China: The political ecology of Chongming Eco-Island. *Political Geography*, *69*, 89–102.

Xie, L., Mauch, C., Tan-Mullins, M., & Cheshmehzangi, A. (2022). Disappearing reeds on Chongming Island: An environmental microhistory of Chinese eco-development. *Environment and Planning E: Nature and Space*, *5*, 225–249.

Xu, J., Yeh, A., & Wu, F. L. (2009). Land Commodification: New Land Development and Politics in China since the Late 1990s. *International Journal of Urban and Regional Research*, *33*, 890–913.

Xu, J., Zou, G., & Chung, C. K. L. (2022). (Re) shaping urban governance through state–business interaction in inland China's emerging industries. *China Review*, *22*, 77–104.

Xue, D., & Wu, F. (2015). Failing entrepreneurial governance: From economic crisis to fiscal crisis in the city of Dongguan, China. *Cities*, *43*, 10–17.

Yang, C. (2014a). Market rebalancing of global production networks in the Post-Washington Consensus globalizing era: Transformation of export-oriented development in China. *Review of International Political Economy*, *21*, 130–156.

Yang, C. (2014b). State-led technological innovation of domestic firms in Shenzhen, China: Evidence from liquid crystal display (LCD) industry. *Cities*, *38*, 1–10.

Yang, R., & Loopmans, M. (2023). Land dispossession, rural gentrification and displacement: Blurring the rural-urban boundary in Chengdu, China. *Journal of Rural Studies*, *97*, 22–33.

Yang, Y.-R., & Chang, C.-H. (2007). An urban regeneration regime in China: A case study of urban redevelopment in Shanghai's Taipingqiao area. *Urban Studies*, *44*, 1809–1826.

Ye, Y., & LeGates, R. (2013). *Coordinating Urban and Rural Development in China: Learning from Chengdu* Chengdu: Edward Elgar Publishing.

Yeh, A. G. O., & Wu, F. (1996). The new land development process and urban development in Chinese cities. *International Journal of Urban and Regional Research*, *20*, 330–353.

Yeh, E. T. (2009). Greening western China: A critical view. *Geoforum*, *40*, 884–894.

Yep, R., Wang, J., & Johnson, T. (2019). *Handbook on Urban Development in China*. Cheltenham, UK: Edward Elgar Publishing.

Yeşilbağ, M. (2020). The state-orchestrated financialization of housing in Turkey. *Housing Policy Debate*, *30*, 533–558.

Yeung, G. (2019a). 'Made in China 2025': The development of a new energy vehicle industry in China. *Area Development and Policy*, *4*, 39–59.

Yeung, H. W.-C. (2017). Rethinking the East Asian developmental state in its historical context: Finance, geopolitics and bureaucracy. *Area Development and Policy*, *2*, 1–23.

Yeung, H. W.-C. (2019b). Rethinking mechanism and process in the geographical analysis of uneven development. *Dialogues in Human Geography* *9*, 226–255.

Yrigoy, I. (2018). State-led financial regulation and representations of spatial fixity: The example of the Spanish real estate sector. *International Journal of Urban and Regional Research*, *42*, 549–611.

Yu, K., Li, D., & Li, N. (2006). The evolution of greenways in China. *Landscape and Urban Planning*, *76*, 223–239.

Yu, Y., Hamnett, C., Ye, Y., & Guo, W. (2021). Top-down intergovernmental relations and power-building from below in China's urban redevelopment: An urban political order perspective. *Land Use Policy*, *109*, 105633.

Yu, Z., & Gibbs, D. (2018). Sustainability transitions and leapfrogging in latecomer cities: The development of solar thermal energy in Dezhou, China. *Regional Studies*, *52*, 68–79.

Yu, Z., & Gibbs, D. (2023). Green industry development and urban sustainability transitions in China's latecomer cities: The case of Dezhou. In F. Zhang & F. Wu (Eds.), *Handbook on China's Urban Environmental Governance* (pp. 355–368). Cheltenham, UK: Edward Elgar.

Yu, Z., & Huang, P. (2020). Local governments' incentives and governing practices in low-carbon transition: A comparative study of solar water heater governance in four Chinese cities. *Cities*, *96*, 102477.

Zeller, C. (2010). The Pharma-biotech Complex and Interconnected Regional Innovation Arenas. *Urban Studies*, *47*, 2867–2894.

Zhan, Y. (2021). Beyond neoliberal urbanism: Assembling fluid gentrification through informal housing upgrading programs in Shenzhen, China. *Cities*, *112*, 103111.

Zhang, B. (2020). Social policies, financial markets and the multi-scalar governance of affordable housing in Toronto. *Urban Studies*, *57*, 2628–2645.

Zhang, F. (2015). Building biotech in Shanghai: A perspective of regional innovation system. *European Planning Studies*, *23*, 2062–2078.

Zhang, F., Chung, C. K. L., Lu, T., & Wu, F. (2021). The role of the local government in China's urban sustainability transition: A case study of Wuxi's solar development. *Cities*, *117*, 103294.

Zhang, F., Chung, C. K. L., & Yin, Z. (2020). Green infrastructure for China's new urbanisation: A case study of greenway development in Maanshan. *Urban Studies*, *57*, 508–524.
Zhang, F., Cooke, P., & Wu, F. (2011). State-sponsored research and development: A case study of China's biotechnology. *Regional Studies*, *45*, 575–595.
Zhang, F., & Wu, F. (2012). 'Fostering Indigenous Innovation Capacities': The development of biotechnology in Shanghai's Zhangjiang High-Tech Park. *Urban Geography*, *33*, 728–755.
Zhang, F., & Wu, F. (2019). Rethinking the city and innovation: A political economic view from China's biotech. *Cities*, *85*, 150–155.
Zhang, F., & Wu, F. (2022a). Financialised urban development: Chinese and (South-) East Asian observations. *Land Use Policy*, *112*, 105813.
Zhang, F., & Wu, F. (2022b). Performing the ecological fix under state entrepreneurialism: A case study of Taihu New Town, China. *Urban Studies*, *59*, 1068–1084.
Zhang, F., & Wu, F. (2023). *Handbook on China's Urban Environmental Governance*. Cheltenham, UK: Edward Elgar.
Zhang, F. and Wu, F. (2024) Green state entrepreneurialism: Building the park city in Chengdu, China. *Transactions in Planning and Urban Research*. https://doi.org/10.1177/27541223241253231
Zhang, F., Wu, F., & Lin, Y. (2022). The socio-ecological fix by multi-scalar states: The development of 'Greenways of Paradise'in Chengdu. *Political Geography*, *98*, 102736.
Zhang, J. (2013). Marketization beyond neoliberalization: A neo-Polanyian perspective on China's transition to a market economy. *Environment and Planning A*, *45*, 1605–1624.
Zhang, J. (2019). The geographic political economy of art district formation in China: The case of Songzhuang. *Geoforum*, *106*, 340–348.
Zhang, J., & Peck, J. (2016). Variegated capitalism, Chinese style: Regional models, multi-scalar constructions. *Regional Studies*, *50*, 52–78.
Zhang, L., & Lan, T. (2023). The new whole state system: Reinventing the Chinese state to promote innovation. *Environment and Planning A: Economy and Space*, *55*, 201–221.
Zhang, L.-Y. (2021). *Conducting and Financing Low-carbon Transitions in China: A Governmentality Perspective*. Cheltenham, UK: Edward Elgar Publishing.
Zhang, L.-Y. (2023). Carbon governmentality in Chinese cities. In F. Zhang & F. Wu (Eds.), *Handbook on China's Urban Environmental Governance* (pp. 128–143). Cheltenham, UK: Edward Elgar.
Zhao, S. X., Zhan, H., Jiang, Y., & Pan, W. (2017). How big is China's real estate bubble and why hasn't it burst yet? *Land Use Policy*, *64*, 153–162.
Zheng, Y., & Huang, Y. (2018). *Market in State: The Political Economy of Domination in China*. Cambridge: Cambridge University Press.
Zhou, X. (2022). *The Logic of Governance in China: An organizational approach*. Cambridge: Cambridge University Press.
Zhou, Y. (2008a). *The Inside Story of China's High-Tech Industry: Making Silicon Valley in Beijing*. New York: Rowman & Littlefield Publishers, Inc.
Zhou, Y. (2008b). Synchronizing export orientation with import substitution: Creating competitive indigenous high-tech companies in China. *World Development*, *36*, 2353–2370.
Zhou, Y., Lazonick, W., & Sun, Y. (2016). *China as an Innovation Nation*. Oxford: Oxford University Press.
Zhou, Y., Lin, G. C., & Zhang, J. (2019). Urban China through the lens of neoliberalism: Is a conceptual twist enough? *Urban Studies*, *56*, 33–43.
Zhou, Y., & Ma, L. J. C. (2000). Economic restructuring and suburbanization in China. *Urban Geography*, *21*, 205–236.
Zhou, Y., & Tong, X. (2003). An innovative region in China: Interaction between multinational corporations and local firms in a high-tech cluster in Beijing. *Economic Geography*, *79*, 129–152.
Zhu, J. (2019). *Urban Development in China Under the Institution of Land Rights*. London: Routledge.
Zhu, J. M. (1999). Local growth coalition: The context and implications of China's gradualist urban land reforms. *International Journal of Urban and Regional Research*, *23*, 534–548.
Zhu, K., Zhang, F., & Wu, F. (2023). Creating a state strategic innovation space: The development of the Zhangjiang Science City in Shanghai. *International Journal of Urban Sciences*, *27*, 599–621.
Zhu, K., Zhang, F., Wu, F., et al. (2024) Governing innovation-driven development under state entrepreneurialism in China. *Cities,* 152, 105194.
Žižek, S. (1999). *The Ticklish Subject: The Absent Centre of Political Ontology*. London: Verso.
Zukin, S. (2020). Seeing like a city: How tech became urban. *Theory and Society*, 1–24.

Index

Pages followed by "n" refer to notes.